Thomas Hill

Geometry and Faith

A Supplement to the Ninth Bridgewater Treatise. Third Edition

Thomas Hill

Geometry and Faith
A Supplement to the Ninth Bridgewater Treatise. Third Edition

ISBN/EAN: 9783337142872

Printed in Europe, USA, Canada, Australia, Japan

Cover: Foto ©Lupo / pixelio.de

More available books at **www.hansebooks.com**

GEOMETRY AND FAITH

A SUPPLEMENT

TO THE

NINTH BRIDGEWATER TREATISE

BY

THOMAS HILL

"The truths of Natural Religion are impressed in indelible characters on every fragment of the material world"

THIRD EDITION GREATLY ENLARGED

BOSTON
LEE AND SHEPARD PUBLISHERS
NEW YORK CHARLES T. DILLINGHAM
1882

"This is the chiefe Glory of Geometry, that it loyters not, or employes it self about these inferiour Machines, from whence it had its Original, but hath soared up into Heaven, and resetled humane minds, (groveling before in the dust) in Cœlestial Seats."—*John Leeke and George Serle,* 1661.

COPYRIGHT, 1882,
BY LEE AND SHEPARD.

All rights reserved.

CONTENTS.

		Page.
CHAP. I.	SYMMETRY IN SPACE	1
II.	SYMMETRY IN TIME	10
III.	NUMBER	15
IV.	THE CALCULUS	20
V.	APPLIED MATHEMATICS	24
VI.	MOTION	27
VII.	MUSCULAR ACTION	32
VIII.	GEOMETRICAL INSTINCTS	37
IX.	MOTION ETERNAL IN DURATION	43
X.	MOTION OMNIPRESENT IN SPACE	46
XI.	THE SPHERE OF HUMAN INFLUENCE	50
XII.	MAGNITUDE	54
XIII.	CHANCE AND AVERAGE	64
XIV.	PHYLLOTAXIS	71
XV.	NUMBER AND PROPORTION	90
XVI.	THE DEVELOPMENT OF FORMS	102

* * * γεωμετρὶαν, * * ὃ δὴ θαῦμα οὐκ ἀνθρώπινον, ἀλλὰ γεγονὸς θεῖον φανερὸν ἂν γίγνοιτο τῷ δυναμένῳ ξυννοιεῖν· * * ὃ δὲ θεῖόν τ᾽ εστὶ καὶ θαυμαστὸν τοῖς ἐγκαθορῶσί τε καὶ διανοουμένοις· — *The Epinomis*.

PREFACE.

"THE present work lays no claim to originality." A humble gleaner in the fields of Mathesis and Theology, I offer only a few of the common fruits, well known to those who have more thoroughly surveyed the boundaries of these two domains. And I have ventured to connect them with the name of the "Bridgewater Treatises," not because I consider myself worthy to appear in company with their writers, but simply that I may thus the more earnestly express my admiration for the treatise of Babbage.

WALTHAM, MASS., 1849.

PREFACE TO THE SECOND EDITION.

I have rewritten the greater part of this work, and altered so much the expression and illustration of my thought, that I might have given it a new title, but for the affection which twenty-five years' familiarity has bred for the old one.

PORTLAND, ME., 1874.

Εὐμολόγητον, ἔφη· τοῦ γὰρ ἀεὶ ὄντος ἡ γεωμετρικὴ γνῶσίς ἐστιν·
Ὁλκὸν ἄρα, ὦ γενναῖε, ψυχῆς πρὸς ἀλήθειαν εἴη ἄν, καὶ ἀπεργαστικὸν
φιλοσόφου διανοίας πρὸς τὸ ἄνω σχεῖν ἃ νῦν κάτω ὂν δέον ἔχομεν·
— *Plato's Republic, Book VII.*

GEOMETRY AND FAITH.

I.

SYMMETRY IN SPACE.

The universe, actual, possible and impossible, is composed of four elements, spirit, matter, space, and time, which are by no alchemy transmutable into each other. Many alchemists continue, even in this closing half of the nineteenth century, to make the attempt, and some even flatter themselves that they are succeeding; but the sturdy reply of human consciousness is, that the four elements are diverse and not transmutable; or, if any transmutation is possible, it must be confined to this, that matter may, in some manner, be an effect of spirit. But to us, finite spirits, nothing more is granted than the re-arrangement, the partial control of matter, not its creation. Matter, as we know it, is distinguished by its being the recipient and dispenser of force; which force, so far as we know it, is from spirit alone. This obedience of matter to spirit gives justification to our suspicion that it is the creation of spirit.

Space and time are without parts, and are indivisible except by a mental act. This division is suggested to us by manifest motion in matter. Force shows itself in matter by moving it; that motion calls our attention to the space and time, within which the motion is taking place; and we divide mentally this space and time, first from the remainder of the boundless contiguities, secondly into smaller parts. Thus geometry and algebra are generated, the sciences which deal respectively with space and time, those pure entities, the relation of which to the Infinite Spirit we cannot comprehend, but which we become familiar with in the finite portion embraced in our experience, in the universe and its history.

In geometry, the mind imposes upon indivisible space arbitrary boundaries of division, according to arbitrarily selected laws or conditions. These boundaries are of three kinds, surfaces, lines and points. The point is a zero of magnitude in space, but nevertheless is not nothing; which is nowhere, while the point is somewhere. This contradiction in terms, that a point should have no extension, and yet have a position, is one of those instances, in which geometry abounds, in which the mind is compelled, by the necessity of direct vision, to admit each of two truths, which are to logic mutual contradictories. The mathematician modifies the law of non-contradiction by confining it to propositions concerning finite quantities.

A lower form of a zero of magnitude in space is the line, which is extended, at each point, only in two opposite directions; and the lowest form is the surface; for which there can, at each point, be drawn a line, such that the surface extends, in every direction, only perpendicular to that line. Geometers define these lower forms of zeroes, or boundaries in space, by the further self-contradiction of imagining the movement of a point; a double contradiction, since space is itself incapable of motion, much more a zero of magnitude.

A geometrical line is defined as the path of a point, moving according to certain conditions, which always limit its motion, in each of its positions, to one of two opposite directions. Or, it may be defined as a continuous series of all the points which fulfill certain conditions, among which must be the condition that each point is contiguous only to two others, one on the opposite side to the other. So also a surface may be defined as the space in which a point moves, when, in each position which it assumes, a straight line may be drawn through it, and its motion be permitted, in any direction at right angles to that line, and in no other. Or, the surface may be defined as a series of points, through any one of which a straight line may be drawn, such that all the contiguous points lie in a direction at right angles to that line. To either of these definitions of a surface, we must add, in order to make a geometrical surface, some other conditions which the points must fulfill.

When the geometer has selected these conditions and would investigate the form which the points, so conditioned, would enclose, he is not contented with the mere act of reason; he endeavors to bring imagination to his aid; to make a sensible image of the form. If he has been blind from his birth, he imagines his fingers feeling out the form; otherwise he embodies it visibly, as in a drawing, or in a model. If he would convey a knowledge of it to others, he calls matter to his aid, and forces atoms of chalk, black lead, wood or thread, to fulfill approximately the conditions which his geometric law imposes upon the series of points. This drawing, or model, is an expression of his idea, an enunciation of his law. A geometrical figure, whether upon the blackboard, or the printed page, or in a block of wood, or a set of stretched threads, is incontrovertible evidence that a geometer has been expressing, by this means, a geometrical thought.

The laws which please the geometer most highly are those

which give us symmetrical figures, figures in which part answers to part; either on opposite sides of one line or one surface, or about more than one line or surface. This taste is not peculiar to the geometer; symmetry pleases the most savage, as it does the civilized man; and men whose whole ability lies in other directions, as well as the mathematician. A striking proof of the universality of this taste was shown in the sudden and universal popularity attained by the kaleidoscope. In a few years that toy of Brewster found its way to every parlor, and the heart of every child, ay, and every man in Christendom. Yet its sole magic consists in the symmetry which it imparts to a few fragments of irregular form. But that magic is sufficient to enchant all who come within its sway. We have never found any one uninterested in an extempore kaleidoscope, made by throwing open the piano, and placing brightly colored articles at one end of the folding lid.

All regularity of form is as truly an expression of thought as a geometrical diagram can be. The particles of matter take the form in obedience to a force which is acting according to an intellectual law, imposing conditions on its exercise. It does not alter the reality of this ultimate dependence of symmetry upon thought, simply to introduce a chain of secondary causes, between the original thinking and the final expression of the thought.

Many of the geometer's *a priori* laws were, indeed, first suggested by the forms of nature. Natural symmetry leads us to investigate, first, the mathematical law which it embodies; then, the mechanical law which embodies it. Thus all the benefits which have come to our race from the pursuit and discovery and use of the keys of physical science, have been bestowed upon us through these suggestions of geometrical thoughts in the outward creation.

But in the pursuit of mathematical knowledge men began, at an early age, to invent and investigate *a priori* laws, laws of which they had not received any suggestion from nature. And the intellectual origin of the forms of nature was made still more manifest when these *a priori* laws, of man's invention, were, in many cases, afterwards discovered to have been truly embodied in the universe from the beginning; as, for example, Plato's conic sections in the forms and orbits of the heavenly bodies, and Euclid's division in extreme and mean ratio.

This division in the extreme and mean ratio was invented by the early geometers, without any known suggestion. It is evident that this division might be illustrated in a great variety of ways. A whole must be divided into two parts, such that the first shall bear the same relation to the second that the second does to the whole. No matter what the whole is, a division of it approximately in this manner would be an expression of the idea of extreme and mean ratio. If the whole were a quantity (distance, angle, surface, volume, value, time, velocity, &c.), and the relation were that of magnitude, the whole would be to the smaller part as unity is to half the difference between three and the square root of five. If, on the other hand, the whole were a work of art of any kind, or a system of thought, the relation would not be one of mere magnitude; and the division would be a work of more ingenuity. But, whatever the whole, or the relation, the proper division would be an expression of the idea.

Now we have, in nature, at least three embodiments of the law of extreme and mean ratio, two of which are very striking. The botanists find that two successive leaves, counting upward on the stem, stand at an angle with each other that is either one-half, one-third, two-fifths, three-eighths of the whole circle; or some higher approximation to this peculiar proportion. The seed vessels and buds on a spike of broad-leaved plantain afford one of the most instructive examples. They are usually set on a high approximation, so that the order is not apparent. Take a piece of the spike, an inch or so in length, between your hands, and gently twisting reduces it to three; while a slight twist in the opposite direction brings out five rows, which a harder twist reduces to two.

The efficient cause of this arrangement we do not know. It has been ingeniously suggested that it might be produced by a simple law of the genesis of cells. Let us suppose that each cell emits a new cell at regularly recurring intervals of time, and that the new cell begins to generate cells at the expiration of two intervals after its birth. A cell developing on a plane, under this law, would produce its cells in the phyllotactic order of the leaves, in the terminal rosette of a plant. But it is difficult to see how this hypothesis can be made to include and explain the whole phenomena of the arrangement.

The final causes, although the devout mind always recognizes the impossibility of man's attaining a certainty concerning all the final causes of the phenomenon, are more obvious. It has been shown that this division of the circle insures in the only perfect way to each leaf its chance at zenith light, its best chance at air; in short, that this phyllotactic law distributes the leaves most evenly about the stem.

In the solar system, if we divide the periodic time of each planet by that of the planet next farthest from the sun, we shall have, beginning with the quotient of Uranus' year divided by that of Neptune and ending with the quotient of Mercury's year divided by that of Venus, a series of fractions agreeing very closely with the approximations of the phyllotactic law. The problem was similar. The planets would not have remained in proper subjection to the sun had they been allowed to group themselves too frequently in one rebellious line, hanging upon the golden chain of his attraction, dragging him and themselves from their proper orbits. They must be kept evenly distributed about the sun; and since they are moving, the times of their revolution, their angular velocities, must be divided by the same law as that which divides the stationary angles of the leaves.

We have then in the plants a geometrical or angular illustration, and in the planets an algebraical or temporal illustration, of the mathematical idea of extreme and mean ratio. The inference seems irresistible,—these two illustrations, which cannot be imagined as having any causal or genetic connection, owe their intellectual relation to having sprung from One Mind.

This is a striking illustration, but the same inference may be drawn from every form in nature,—planet, crystal, plant, and animals. All natural forms conform more or less closely to geometrical ideals; sufficiently near to suggest these ideals to men fitted to receive the suggestion; sufficiently near to show that the whole of nature may, in one sense, be regarded as a series of drawings and models, by which to teach the mathematics to students in the school of life.

The final causes may never, however, be considered as wholly known. The perfection of the Divine workmanship is shown in the adaptation of each object in nature to a great variety of ends. The geometrical laws, on which the world is built, are adapted to all the wants and all the needs of every creature. Our human needs are innumerably various, and nature finds means to satisfy them all. Our intellect craves symmetry, and through symmetry is first led to the perception of geometric law. But we love the symmetry before we perceive the law. The sense of beauty is satisfied, even in externals, most perfectly, and fills us with most pleasure, in things that the understanding fails to analyze and define. Much has been written concerning an analysis of the beauty of outline; one great painter thinking it consists in flexure, others assigning it to a spiral, or a helix, or an ellipse; while Darwin refers it to early association, while yet a suckling, with the form of the mother's breast. I venture with diffidence to give my own opinion, that the perception of beauty in outline is the unconscious perception of geometric law, — just as the perception of harmony has been demonstrated to be the unconscious perception of arithmetical ratios in time, or algebraic law. The beauty of outline, I would say of external form, independently of expression, is in proportion to the simplicity of the geometric law, and to the variety of the outline which embodies it. ' Nor is it essential to the highest enjoyment of beauty that the conformity to geometric ideals should be perfect, any more than it is essential to the highest music to have the harmony perfect. On the contrary, the higher degrees of beauty are apt to be found in forms that suggest, rather than embody. the ideal; and especially in figures potentially, but not actually, symmetrical. The monotony, which might result from unbroken regularity of form, is avoided, and a new grace is given, for example, to the higher animals, by their temporary disguise of symmetry. in their varied positions and movements. In the sea shells, the same end is attained by the spiral form, which so many of them take; in which there is not an actual symmetry, but only a law of symmetry, the perfect development of which would require an infinite number of convolutions.

In the forms of vegetative life, there is the widest departure from actual symmetry, and yet a constant suggestion of its laws.

The phyllotactic law secures to the tree a general regularity, and equal growth upon every side ; and yet, by complication of detail, combined with occasional failure or destruction of buds, secures an endless variety of graceful forms, in each species. May we not then name beauty as another final cause, another end secured by the adoption of the division in extreme and mean ratio ? The approximations are beautiful to us, and the pleasure given to us was foreseen when the law was adopted. May it not also have been felt ; and may not the forms of flowers be but approximations toward the expression of an infinite beauty, hidden, from all finite sense, in the incommensurable ratio of that surd ? That the external symmetry of animals may have beauty as its final cause, is rendered probable from the lack of symmetry in the viscera, which are hidden from sight.

Whatever be our speculations upon such points, this at least is manifest, that the sense and the presence of beauty are kindly adapted to each other in the world. Even shapeless matter declares its Creator's power ; the perfect symmetry of crystalline forms, the potential symmetry of all the organic worlds, show forth **His wisdom** and **His love.**

II.

SYMMETRY IN TIME.

TIME has but one dimension, and is divisible only into before and after. In the zero of now, the future is becoming the past; and this suggests the division of the future, and of the past, by the insertion of imagined presents, zero boundaries, dividing time into periods; these imagined future nows becoming actual, as we successively reach them, and those past having been actual, as we passed them. As time flows only forward, the imagination runs backward into the past with the greatest difficulty; indeed I am not certain whether it is possible for the imagination to run back; when I attempt to do so, I find that I leap backward, by longer or shorter leaps, but never run continuously in imagination through time, except forward, from the moment to which I have leaped.

By symmetry in time, therefore, we do not mean a similar arrangement of intervals before and after a certain moment. This has occasionally been attempted in *per recte et retro* chanting; but it is a transference of geometrical symmetry to time, where it is out of place, and tasteless. Symmetry in time is the arrangement of two or more similar series of intervals, to follow the same, or successive movements. When the set of similar intervals follows the same moment, it constitutes keeping of time; when successive moments, rhythm; unless the intervals are very short, when rhythm becomes tone, or color, and keeping time becomes harmony.

SYMMETRY IN TIME. 11

The passion for harmony and rhythm is an essential element in human nature. It is a passion which varies greatly in intensity in different persons, but it is never wholly absent. Savage nations have some rudiments of music and of song. The naked Fuegian, when the stormy winds of his inhospitable straits pause for a while in their wild uproar, chants his songs or hymns in a rude measure and melody. The dullest ear for harmony has an ear for rhythm sufficient to perceive the difference between prose and verse. It is when the intervals of time in rhythm become so short, as to be separately imperceptible, that the rhythm is called simply tone; and harmony is the simultaneous movement of tones. Man finds pleasure in all forms of symmetry in time, whether the parts are perceptible, or imperceptibly short; and the world has been made in exquisite adaptation to this taste of man.

Space has three dimensions, time but one. Yet, in some respects, time is richer in its contents for man, than is space. The beauty of forms in space, is almost equalled by the beauty of color, and color arises simply from symmetry of times; it is a kind of tone. Color, indeed, is more expressive, more directly productive of pleasure to the eye, than form. The latter appeals more to the intellect, and is more directly expressive of intellectual ideas; the former appeals more to the heart, and gives a sweeter pledge of the Divine Love. But beside color, symmetry in time gives us music in all its vast variety of forms and expressions. In music there is a beauty as distinctly intellectual as that of geometrical figures, and a power of expression which geometric form attains, scarcely even in the human figure and face. Nor can we omit to mention heat, which although not giving direct pleasure to the mind, and the heart, as beauty, color, and music do, is still essential to the life of the body and to its comfort. Many chemical changes are also produced by minute symmetric motions.

A minute symmetrical division in space produces no sensation for us, except as it may lead to symmetrical motion. Thus the minute symmetry of the particles of a solid in a clear liquid solution, is revealed to us only through the motion of light, and the changes which that motion experiences in going through the solution. In like manner the temporal symmetry of motion in the ray of light coming from the stars, brings us more information than its mere property of making visible can give; for, on cross-examination by the spectroscope, it confesses the chemical secrets, and the degree of heat in the star at the time of its leaving, ages ago; and also the direction of the star's motion.

Man's organization, and his surroundings, are adapted to his love of music. His voice is capable of being regulated to musical tones of various pitch. The metals are sufficiently elastic to render them sonorous; and, as the foundation of all, the air itself, by its elasticity, becomes the vehicle of sound and the instrument of music.

Two gases, intermingled, remain to a certain extent independent of each other; and, inasmuch as sound travels in each gas with a velocity proportioned to the density and elasticity of that gas, there will be, from a single source of sound, two sounds propagated in the mixture, with different velocities, interfering with each other, and destroying the pure tone of a musical sound. Now the atmosphere is a mixture of heavy oxygen, with lighter nitrogen. The elasticities are, however, so nearly adjusted to the densities, that sounds travel in either gas with nearly the same velocity, so that the air sounds in an organ pipe, as if one gas. Had sound travelled in these two gases at rates differing as much, as the rate in them differs from that in most of the gases known to us, the use of wind instruments of music would have been impossible; probably all music, even the tones of the human voice, would, in that case, have been discordant to an ear at any considerable distance from the source of sound. With the intense and elevating character of the pleasure derived, first from the tones of human speech, from the melody of birds, and other natural music, and secondly from the art of music, in our minds, we cannot but be grateful for this adaptation of the mingled atmosphere to the needs of man, in his higher nature.

All undulatory motion produces a symmetrical division of time. The beauty of color, like that of tone, arises from an implicit perception of rhythm. The harmony of tints in the landscape, like that of the sounds in a strain of music, arises from the harmony of times in which the vibrations of the mediums occur. The pleasure, in either case, arises from an implicit, or unconscious, perception of keeping time. Heat also has its colors, or tones, as is known to all who have noticed that the sun's heat passes freely through glass; which is impervious to the heat of a fire. What other advantages to man may hereafter be discovered, in this coloration of heat, time alone can show; but when we consider to what an extent, through the providence of God, glass is employed, it seems not irreverent to own our gratitude to Him that this substance reflects back the warmth of our apartments, and keeps it within, but allows the heat of the sun to pass through from without.

Besides these hidden proofs of creative foresight, and beneficence, in the concealed, or minute symmetry of time, we shall find open and abundant proofs in manifest rhythmic movements. In the play of alternating muscles, we perceive an adaptation of the physical frame to the intellectual taste. In walking, for example, there are few persons who do not feel the increase of pleasure and of power gained by keeping step. A single drum-tap, regulating the tramp of a large body of men, has sometimes an effect almost equal to that of music. The rhythm of verse, and of music, delights many who are comparatively insensible to both melody and harmony properly so called. Who that reflects upon the genius of Bach, of Handel, of Haydn, and Beethoven, and considers the effect which music, such as theirs, has upon the world, can doubt the kindness of that superintending Power, who kindled the fire in their hearts, and through that in ours; who also adapted the air, and the various materials, for man, by which he pours out his musical conceptions? Who that reflects upon the genius of a Homer, and a Shakespeare, and remembers to how many millions their verse has given delight and instruction, can doubt that the same Beneficence gave the poet his power, and men the heart to be touched with **poetry**?

III.

NUMBER.

NUMBER is not, like space and time, matter and spirit, an integral part of the Universe, nor is it a necessary attribute of either of these. Space and time are without parts or limits, and are, in themselves, so diverse that they would not suggest even the idea of duality. Thus also amorphous matter suggests no number. Number is an impress of thought, it is a pure creation of Spirit; and its constant suggestion in the forms and periods of nature, is a clear demonstration that nature is the work of an Intellect which controls both space and time in thought. The human intellect early learns number from the text-book of outward nature; and delights in tracing, further than nature goes, the laws of number.

From the great usefulness of this earliest abstract science, and from the fascination of its pursuit, arithmetic has, in modern schools, been allowed to usurp the place of geometry; and the pupil has been taught to reason upon abstract numbers before he has learned to conceive clearly imaginary forms. From the same fascinating power, number has sometimes, in the minds of great men, like Pythagoras, been allowed to occupy a disproportionate share of attention; as though number included all proportion and beauty. Even the Hebrews, with all their clearer light of truth, appear to attach a mysterious power to number.

There is a power in number. When our human thought attempts the survey of space and time, and would subdue these realms to obedience under our intellect, we find ourselves compelled, before we can attain any precision in our forms, to introduce number. The reason can deal, to some extent, with continuous quantity, moving under continuous law, and not in the proportion of numbers. But the imagination cannot take a step with any clearness, much less can the hand build with any satisfaction, without referring quantities to a unit of quantity, to which the ratio shall be that of two numbers to each other. And of course our finite intellects handle with most ease the smaller numbers; so that these become to us the most important; and there is not a number under ten which has not some strong associations with it in the human mind, which give it a kind of sanctity. These mystic charms cluster especially around the odd numbers three, five, seven, and nine; which seem to have an individuality; the first-named three being primes, while four is but two twos; and six, two threes; and these charms were felt in the earliest ages of human history.

But nature also loves these numbers; and they are illustrated, even to the untaught mind, by many phenomena; organic beings possess a unity, which is absolute; the sexes, of both plants and animals, give us duality; the powers within, and those above, suggest the threefold division; the points of the compass, the limbs of mammals, give us the number four; the fingers of the hand, five, and so on. And the increasing knowledge of the physical world, in our nineteenth century, brings us increasing proof that God, who planned heaven and earth, was acquainted with numbers; made all things in number, weight, and measure; and adopted the smaller numbers, either out of preference for them, or in condescension to the minds of his children, whom he has placed here for their preparatory education.

Chemistry is a science of this century, and it teaches us that, from the beginning, the numbers two and three have been dominant powers in the Universe. Simple unites with simple to form a couple, a compound. This couple rarely takes a third element to form a triplet. The couplets and triplets unite again in compound couplets, and thus the innumerable variety of substances is built up under the simplest possible combinations of number. Follow these substances through all their various modes of motion and action; in their weights, in their attractions, their gaseous condition, their volume, their specific heat, their color at a high temperature; and they are found still to be bound together by simple laws of arithmetical proportion.

Consider also the law of extreme and mean ratio, as exhibited in the leaves of plants. In itself the law transcends the power of number, and had the plants fulfilled it with absolute accuracy, it might have been, even yet, hidden from the mathematician's eye. But the plants give it to us only by approximations; approximations which demonstrate that the exact law was known to the Builder of the plant, and is by him revealed to the mathematician; but which give to the unlearned the simpler conception of the first four prime numbers; in the beautiful varieties of leaves opposite, and leaves ternate, five pointed and seven pointed stars.

The laws of musical harmony are especially to be noted. When the waves of the air are perceived only as continuous musical tones, and the individual vibrations are not at all recognized, why should the ratio of four to five give us pleasure, and that of eight to eleven give us none? What process of education in our ancestry, what association of ideas, renders the effect of the one combination harmonious, of the other discordant? Any attempt to explain it will but strengthen the conclusion, that to the Builder of the ear the laws of number were known, and that the ear was constructed with reference to them.

The harmonies of light and heat are not sufficiently well understood to make the argument here so apparent. Yet there is, doubtless, in these departments also, an adaptation of the human sense to the perception of effects arising from simple numerical proportions in the frequency of vibrations. In the matter of geometric form, while the value of proportion has been felt by all artists, and all architects, the value of numbers in the proportions has not been universally conceded, nor its place assigned. Yet I have by experiments, upon unprejudiced persons of good taste, strengthened greatly my inclination to accept Hay's law, — that angles, real or potential, are the essential elements of geometric beauty; and are beautiful in proportion to the numerical simplicity of their ratio to the right angle.

With these manifest indications that the divine thought, the ideals of the creation, include number as an essential element, we may well understand the enthusiasm of early thinkers over the properties of the smaller numbers. The sacredness of the number three has been made especially prominent in Christendom. The four elements of the ancients, and Erigena's fourfold division of nature, show the power of the points of the compass to impress their number on the human mind. The five digits of the hand, and the prevalence of fivefold divisions in the floral kingdom, give us the five-pointed star with its symbolism; point up, for manhood and virtue; point down, for beastliness and sin. The lily tribe gives us the six-pointed star; and six, a perfect number, in which the sum of the factors equals the product, is fitting as a symbol of the descent of the divine into the human trinity, the indwelling of God in man; the Perfect perfecting his child. The seven notes of the diatonic scale, the seven distinct colors, and other natural examples, fall in with the seven days of the week, the quartering of the moon's period. Jew and Gentile alike have hallowed the number seven, and no other number occurs so frequently with sacred associations in Jewish and Christian literature. Higher primes than seven do not enter much into our human thought, nor appear to be embodied distinctly in any part of creation known to us. The weeks in the year are four times thirteen; that is, there are about thirteen moons in the year; the only example I remember of a prime number above seven prominently suggested by nature. The nine muses, the ten numerals, the twelve months, and twelve apostles are numbers not prime.

Music, painting, the coloring of nature and art; architecture, sculpture, drawing, the beauty of proportion and form; how large a portion of our earthly pleasure and spiritual culture depends on these; and these draw their charm in some mysterious way from the numbers two, three, five, seven. The number of prime numbers is unlimited; and since the first four give us, in the harmony of tones and colors, and in the proportions of form, such varied sources of high pleasure, such potent modes of spiritual expression, we may reverently hope, that in the immortal life, the same Beneficent Power which makes two, three, five, and seven, thus minister to our joys below, will open to us more of the infinite treasures which lie hidden in the boundless fields beyond.

IV.

THE CALCULUS.

Space and time are so entirely diverse in their nature, that there is no connection or relation between them; except through the mind, as percipient of both; or through will, manifesting itself in motion. In contemplating space we see it as external to the mind; our consciousness does not sharply locate its own whereabouts; we fancy ourselves near the Eyegate or Eargate of the town of Mansoul; but cannot say precisely where our council chamber may be situated. Not so with time, our consciousness is sharply defined; we are neither in the past nor in the future; our conscious moment is the now, without duration. Hence we can more readily imagine ourselves freed from limitations of space than from those of time. We can imagine to ourselves time in the flow of our own thoughts; the thought of space necessarily takes us out of ourselves. But when we go out of ourselves and contemplate space, we carry time with us in the very action of our thought. In all closer contemplation of outlines, the attention is transferred successively to different points of the figure, and time is occupied by that transfer. Thus we come naturally, and almost inevitably, to regard the line as the path of a moving point, the surface as generated by a moving line.

Thus space and time, though heterogeneous, are united into one science of mathematics by human thought; and the laws of algebra, or time, are applied to geometry, or space. By this simple device, into which Descartes and Newton were led by nature's own guidance, the human mind has extended almost indefinitely its geometrical acquisitions; it was by carrying, as it were, its native element of time with it into the domain of space that it has conquered so vast a field.

When we remember how intense the delight which man feels in the discovery of mathematical truths; how many of the noblest thinkers of the race have owed their finest discipline to this pursuit; how rich the harvest of practical benefits which have flowed from the application of mathematics to the arts and sciences; how magical their effect has been in banishing superstition, and elevating the general tone of human thought and human endeavor, we may surely own, with gratitude, the marks of divine wisdom and love, in this gift to man, of the power to penetrate space, and apply to it the laws of time. It is a peculiar gift, not a necessary accompaniment of intellect, for sometimes the brightest intellects possess it in only a very feeble degree. Thankfully, therefore, do we acknowledge the presence of an Infinite Spirit, giving good gifts to man in the inspiration of a Leibnitz and a Lagrange, as well as of a Handel and a Shakespeare.

The main source of this power given by algebra to the geometer, is the comprehensiveness of the language put into his hands. The introduction of general and abstract terms is always a means of enlarging the grasp of thought, and increasing the clearness of reasoning. Space has its three dimensions, its elements of magnitude and direction; and although, in one aspect, the simplest of all possible objects of thought, may yet, for purposes of reasoning concerning it, be advantageously reduced, by algebraical language, to the one term of quantity, capable only of flowing in one direction, and being considered as greater or less than a given magnitude. But the generality thus introduced is made vastly more general by using symbols which shall combine, in one letter, various forms and relations in space, defined according to judiciously selected and easily interpreted laws. Thus, for example, all possible triangles, plane and spherical, and all their properties are implied in the single equation, $r = pq$; and a similar condensation of meaning is attained in mechanical science. Another source of the peculiar power of the calculus arises from the plasticity which it gives to infinitely rigid space. In experimenting upon a rectangular beam, cut from a round piece of timber, we can readily determine its strength when set edgewise; but cannot tell what the strength would have been had the sides been in different proportions. The rectangular parallelopiped inscribed in a cylinder is as absolutely fixed in its dimensions as the hewn timber, but by expressing those dimensions in language borrowed from the science of time, we can imagine them changing in their proportions, and the strength changing with them. Thus we can determine the precise proportion they must bear in order to give the strongest possible rectangular beam that could be cut from a round log. This illustrates, by a simple example, the power given to geometry by Newton's conception of fluxions, his introduction of the idea of velocity into the consideration of form.

The appearance of the same algebraic law in the creation, under the two forms of time and space, has already been alluded to as proof of unity of design; the angles of leaves and the angular velocity of planets being expressed by the same series of fractions. Other examples confirm the sublime induction. The elasticities of gases, strings, and rods are so fundamentally different in kind that we see no connection between them. The elastic force of the stretched string we need not determine; that of the rod, and that of the gas, can be determined only by experiment, and when determined they have no very apparent connection or relation with each other. Nevertheless, each of the three has a peculiar relation to the force of gravity; of which it is, nevertheless, entirely independent. The velocity of a sound traveling in the air, near the earth, would be, were no heat developed in the action, equal to the velocity acquired by a body falling from a height equal to that which the atmosphere would have could it be all compressed to the density of that near the earth's surface. The velocity of a wave traveling on a string is equal to that which would be acquired by a body falling from a height measured by the length of the same cord equal in weight to the tension of the string. And if we take a very fine glass thread by its two ends, the infinitely varied and beautiful forms which it can be made to assume, of waves and folds and kinks and loops, the figure eight and the circle, are all expressed in mathematical language by the same forms as those which express the motions of an ordinary pendulum, under the forces of gravity. The genetic connection, between these forms and these motions, we do not see, any more than that between the times of the planets and the angles of the leaves, but the intellectual connection we detect, and it leads us to recognize with reverential awe the presence of Intellect in the disposition of the particles of both gaseous and solid bodies.

V.

APPLIED MATHEMATICS.

Perfect symmetry belongs only to the ideal, not to the actual. The algebraic conditions are exactly fulfilled by points of space, in an invisible and eternal reality; to this real form, conforming to the algebraic ideal, the material embodiment makes at least a rude approximation. The algebraist devises conditions of various degrees of complexity, delighting chiefly in the simplest; and especially in those giving, with simplicity of conditions, the greatest variety of resulting forms.

The symmetrical forms of nature suggest to man the invention of laws of symmetry, at first simply to explain nature, then to anticipate her work; leading to new examinations of that work. Thus the great mathematical sciences have been alternately the creation and the creators of physical science. The physicists have been prone to deny that the mathematics constitute a science; they have inclined to pronounce them only a key to science, a convenient language wherein to discuss the problems of matter and motion. The mathematicians, on the other hand, whom we should naturally consider the best judges of what their own work is, have declared that geometry is the science of space, algebra the science of time, and that these are simply the first subjects handled by the human intellect with sufficient freedom, vigor, and precision to enable us to draw necessary conclusions. As for geometry and algebra being mere keys to physics, the mathematician would sooner declare the whole visible creation a mere set of models and diagrams wherewith to illustrate the laws of space and time. Whichever of these conflicting views is right, it is unquestionable that the highway to the temple of truth leads alternately from mathematics to physics. Observation alone can lead to nothing, without insight, — without that clearness of inward vision which sees more than the outward fact, sees the divine ideal which the fact partially embodies.

Now in this sublime ascent to knowledge the first steps are easiest, and the way to them has been made exceedingly plain and attractive. "In the beginning the Creating Spirit embodied, in the material universe, those laws and forms of motion which were best adapted to the instruction and development of the created intellect." The circle and the ellipse are among the simplest of figures, defined by the simplest laws. Accordingly the Creator has strewn examples of the circular form around us on every side; and, by the pictured alphabet of the heavens, called our attention to the consideration of elliptical orbits. When, in the course of ages, some of the comparatively easy problems of astronomy had been successfully solved, problems of more difficulty were gradually brought into view; and phenomena which were not obvious, not pictured alphabet, but the fine print of creation, led men into the hidden knowledge of optics, electricity, chemistry, and other forms of molecular physics. The course of history and of scientific progress has been precisely what it might have been had God designed to educate men; to reason with them and teach them the sciences; for there has been a constant presentation of simpler truths, whereby men have been led to the acknowledgment of those less obvious; and this is essentially reasoning.

Four centuries before the Christian era, the mathematicians of Greece were lured into the study of the conic sections; and this prepared the way for the mathematicians of later ages to discuss fully certain equations of the second degree. These were sufficient for all the more obvious phenomena of astronomy and mechanics; and as the demand for higher mathematics has been made by physics, the supply has been granted. The faith, which prompts the scientific investigator to his labor, he may never have expressed in words, but his actions show us what it is, — an inborn, ineradicable conviction that the outward universe is intelligible, and shall at some day be understood. But that day ever recedes, into the glorious future, as we approach it; the rate of scientific progress increases from decade to decade, and yet the new problems, and the new instruments for their solution, increase more rapidly. The Divine Intellect can never be exhausted by the human.

A more detailed examination of the history of the separate sciences would only confirm our conclusion, that, in the selection of laws under which to subject the universe, God has chosen, for those things which would first press themselves upon man's attention, those which are most readily interpreted by man's intellect; and employed more intricate laws for things which would naturally escape man's notice until the state of mathematical science enabled him to take higher problems; in which we recognize evidence of that kindness and foresight, that care for our education and our growth in knowledge and wisdom, which is an inspiring pledge to us that we are indeed children of the Most High.

VI.

MOTION.

The universe about us is in motion. Nothing on which the eye can fall, or the existence of which the hand of science can demonstrate, is at rest. The sun rises and sets, the moon waxes and wanes, the very stars are in motion, to the telescopic eye. Clouds drive over the heavens, and billows roll over the deep; vapors rise from the ocean, rivers run to the sea, and the free winds play around the globe. Plants are ever growing or decaying; and animals maintain their waste, or their waste consumes them. Our modern theories show that the sensible properties in inanimate and apparently motionless matter, such as temperature, color, weight, are really modes of motion in the particles of matter; and this re-echoes the sublime statement of the earliest seer, that the introduction of motion into the universe was the first act of creation.

For, upon a closer examination of motion, and more accurate investigation of its laws, what do we find? That the first law of motion is this: A body, free from external influence, moves with uniform velocity in a straight line forever. This is the first law of motion, derived from the widest generalizations, by legitimate induction from observations, on an immense variety of motions, in nature, and in the laboratory. But to what an astonishing result does this law lead us when we apply it to the case of a body at rest, the velocity of which is nothing. A body at rest, free from external influence, would remain at rest forever. In other words, the first result from the scientific observation of motion in matter is, that matter cannot move. Hence follows the inevitable conclusion, that the cause of all the motion in the universe, is something else than matter. Higher than this the investigation of motion itself cannot lead us; but this is high enough for a most valuable stepping-stone.

Why do we ask the cause of motion? Whence do we derive the idea that there is a cause for it? It is not simply the impossibility of our imagining a beginning; the beginning of motion we often see. But the motions which we most narrowly examine are those produced by our own will; we are conscious that our own volition is the cause of such motions; and this consciousness is the foundation of our faith that motion always has a cause. Is this foundation trustworthy? Beyond all question it is. Nay, it is the foundation of all possible physical science; no man can extend a generalization beyond the particular instances for which he drew it, unless he leans on this consciousness of causing. To return to motion, — matter cannot move, our will can move it; there is nothing to suggest any other origin for motion than volition; hence we naturally, and legitimately, infer that the motion which we see, everywhere in the universe, is produced by a will, independent of matter, and superior to all the phenomena.

Thus the first law of motion, established in the earliest revival of science, demonstrates not only the existence of God, but his perpetual presence and action. Every moving thing in the heavens, or on the earth, bears the same sort of testimony to his being and presence, as that borne by the human voice and action to the presence of a man. Whenever we see anything in motion, God is the mover. In the ancient tongues this was one of his names. The winds blow at his command, the sun rises because it is his will, the falling rain and running stream are his gift; and each beating pulse, each breath that we unconsciously draw, is a proof that this machine of the body is, each moment, dependent on the sustaining love and power of its Creator.

Since we thus refer all motion, even that in our own frames, to the will of God, it may be thought that we are destroying man's freedom, — making him a mere machine, kept in motion by the Maker's supervision. But this objection to the doctrine of man's present dependence, forgets that the consciousness of our freedom is the very basis on which we have built our faith in the existence of God. It is from our own consciousness of power, to cause motion at our own will, that when the first law of motion has excluded us from ascribing it to powers inherent in matter itself, we ascribe all motion to his will, rather than to any unconscious natures.

This consciousness of our own power, our own will, may be denied in words; but it will presently betray itself, lurking in the mind; it cannot be really denied; it is the foundation of all philosophy and faith. The body, in all its molecular changes, by which ultimately the free movement of the limbs is produced, is moved by the will and power of God; the first law of motion proves that; yet the direction of the movement in the limbs is with man, consciousness testifies directly to this; man is free, and cannot heartily believe himself to be otherwise.

Our muscular power is not ours, but it is, to a certain extent, under our control. We cannot lift a finger without the aid of him who formed us; and yet it is we who move our hands. So the engineman, who has not, in his own muscles, strength to drive a single loom, yet, by controlling the valves of his engine, keeps the machinery of many spindles and looms in motion. Thus, with all man's frailty, and his absolute dependence upon other powers, he yet remains a cause, — free and efficient to control and direct the engine of his body, wonderfully framed and intrusted to his care.

Our argument has been, thus far, drawn only from the uniform velocity of motion; but the second clause in the first law would lead to the same result. A moving body, free from external influence, moves, not only with uniform velocity, but in a straight line forever.

As we have no apparent examples in nature of a uniform velocity, so we have none of uniform direction. External influences perpetually accelerate or retain the velocity and change the direction of moving bodies. But as the first part of the law is derived, not from actual examples or instances, yet by irresistible induction from observed facts, so the second part follows by like unavoidable inference from phenomena; indeed, both parts are defended, by some mechanicians, as axioms, needing no other proof than Leibnitz's principle of the sufficient reason.

As the varied motions of the universe cannot have sprung from the action of matter, that being inert, so the constant changes of direction in the motions prove that the forces, independent of matter, are still acting. The rebounding of a solid from a solid shows that the particles of the solid adhere by some form of force different from a cohesion of contact, — elasticity implies that the particles are held together by some force which permits their distances from each other to vary within certain limits. When the ball leaves the muzzle of the gun its path instantly begins to be concave toward the earth; and would be so at any conceivable degree of velocity. The meteor passing the earth at eighty miles a second bows to her as he passes. Thus the moon also is perpetually deflected from its path by the earth, and the earth by the moon, and both are turned constantly aside from their straight course by the sun; and the whole host of heaven is constantly moving in a rhythmic dance, wherein each star influences the motions of the whole, and is influenced by the movements of each of the others.

Our consciousness that we cause motion leads us to ascribe all change of velocity to force, all force to will. The same consciousness bears witness also that all change of direction implies the influence of will. The weight of bodies, the attraction of gravitation, the correlated forces of the universe, these are but reverent forms of words in which we speak of that which can only be referred to the Divine Will. The untaught man, the poets of the earlier ages, were more true to reality when they used more religious forms of speech. It is not so much figurative, as literally true, to say that He who formed the Seven Stars and Orion still guides them on their way. Their circling orbits by their figure, and the golden orbs themselves by their motion, continually manifest the presence of His guiding hand. The forces of cohesion and repulsion, of electrical and chemical change, of heat, of light, — all of the forces by which the existence of a particle of matter can possibly make itself known to our human senses, are but manifestations of the living action of the Most High.

Thus the first law of motion leads us to see God in all things, and all things as the present creations of his hand. It might lead us astray, it might lead us to Pantheism, were it not that it first leads us to perceive that force is an attribute of will, and independent of matter; thus keeping us to the conclusion that the Creator and Governor of all things is free, living, — and our hearts add, good.

VII.

MUSCULAR ACTION.

We have spoken of the human frame as an engine of wonderful construction, whose movements are made dependent on the human will. Yet it is manifest that more of its motions are independent of the will than are dependent upon it. The involuntary muscles, and the involuntary movements not muscular, are those which are essential to the very existence of the body. The circulation of the blood, and its purification through the alternating expansion and contraction of the chest, are obvious instances of these vital, involuntary actions. Not less important is each one of a thousand hidden operations, — capillary movements, glandular secretions, the removal of the effete and the replacing of the living molecules; to say nothing of more muscular actions, — the peristaltic motions, and the wonderful unconscious artifices of swallowing, coughing, sneezing, and the like.

The voluntary muscles are also capable of involuntary action. This is shown not only by occasional convulsive twitchings, or more violent convulsions, but, in a still more instructive manner, by inveterate habit, operating in sleep, or even when the will opposes. But, although the action which has become habitual is not done from distinct, conscious volition, the habit is originally formed by acts in obedience to the will. The law by which the voluntary action becomes involuntary habit, although it reduces, too frequently, man to slavery, is truly beneficent in its design, and in its best effects. The simplest statement of the law would, perhaps, be found in saying that actions which have been associated in volition become associated in execution. In other words, when we have done several things at the same time, or in quick succession, the attempt to repeat one of these actions will tend to produce an involuntary repetition of the others. For an illustration of the beneficent action of this law, we may take the skillful player upon a musical instrument, who is conscious of a volition only at the commencement of each musical phrase: the fingering of the separate notes comes from associated execution. In a familiar piece his volitions would be even less frequent, being necessary only at the commencement of a new strain.

This case of the skillful player, being less frequent, seems the more striking; but there is scarcely an action in life which is not aided by the beneficent operation of the same law, just as there is scarcely a mental action which does not illustrate the kindred law of the association of ideas. The child, just learning to walk, makes a painful effort at each successive movement of each muscle called into action. It requires all the concentrated energy of his will to make the successive volitions necessary for simple stepping from chair to chair. But in a few months he is able, by associated execution, to set in action, by a single volition, a series of alternate motions, that carry him forward, without his attention, in a given course, at a uniform speed. No power of will is required in walking, except when we wish to alter the velocity, or the direction, of our movement.

When the successive movements, dependent on associated execution, are connected, as in walking, by a law of simple alternation, the case is not difficult of explanation; and the physiologists show us how the will relieves itself from duty by a switch, turning off the currents of sensation and command from entering the main office in the brain.

Other cases, in which the operation of the law is no less important to our comfort and convenience, require, however, much more intricate combinations of movement. In many familiar occupations we require our hands to guide an instrument rapidly and freely through outlines of complex, but definite, forms; as, for example, in writing, or in free-hand drawing. All men have more or less of this power to execute ideal figures, or to imitate given forms. This power has been gained, like that of walking, only through repeated and laborious efforts. When moving the hand in one direction, we need a new volition to change its direction, or to alter its velocity; hence our first attempts at curvilinear motion produce polygonal lines. In order to produce a curve, as in the ordinary forms of the capital letters, we must produce, by several muscles acting at once, motions in several directions at the same time, each movement varying in velocity according to definite laws. To draw, for example, a circle, by any conceivable set of muscles operating on the arm, would require at least three sets of muscles, each acting in a different direction, and no two exactly opposed; and these would be obliged to accelerate and retard their action by peculiar laws. The circle, however, is the simplest of all curves: in the ordinary operations of writing and drawing, the rates of acceleration and retardation must follow more complicated laws. Of these laws we think nothing, we know nothing: we see the curve which we would form, and a single impulse of the will sends the pencil along the waving outline.

Charles Babbage, a successor of Sir Isaac Newton in the Lucasian chair, has won an immortality of fame by inventing a machine which will tabulate in numbers the results of any algebraic law which it may be set to obey. But how much more wonderful is this calculating engine of the human body, which is not confined to arithmetical results, nor does it require that its director should be learned in algebraical notation to set it at its appointed task, but which is set by the artist, with his delicate perception of the beauty of form, to embody his divine ideal! and it obeys, and places before us on the canvas, those figures, which, unconsciously fulfilling algebraic or numerical law, reach far higher, and express the spiritual thoughts and purposes of the Master. Is not the Maker of this wondrous engine of the human body worthy of gratitude and adoration?

Consider how wonderful is the phenomenon of a boy's throwing, successfully, at a mark. The epicycloidal theories of Hipparchus, the Newtonian theory of gravitation, the resolution of centripetal and centrifugal forces, the conic sections of Apollonius, the modifications of those curves by the resistance of the air, — all these are involved in the problem, and must be practically solved, with considerable accuracy, before the school-boy can give his fellow a good ball, or catch one on the fly.

It may be observed that the mechanical contrivance by which the human hand is enabled to go through all imaginable motions, and strike, at a free sweep, any curve, however complicated or however beautiful, is an embodiment of one of the most celebrated of mathematical conceptions, discussed in the writings of Plato and Aristotle, constituting, in its development, one of the chief triumphs of Hipparchus, and brought by modern mathematicians, through the arithmetic of sines, and the *canon mirificus* of Napier, into a form capable of reducing to a regular curve the most variable and irregular table of observations. In this method of epicycles, as used by the modern computer, a series of arms is supposed to be carried, each on the extremity of the preceding, and, during the revolution of the first, each to revolve once oftener than the preceding; that is, while the first arm of the series revolves once, the second revolves twice, the third three times, and so on. It only remains for the computer to fix the length of these arms, and determine their original position, in order to make the end of the third or fourth, or, in cases of difficulty, of the fifth and sixth, describe any path he wishes. In the human limb, the upper arm is the first, the fore-arm the second, the hand the third, and the fingers the fourth, fifth, and sixth of these rotating arms; and the fixedness in the ratio of their length is more than compensated for, by our ability to graduate the ratio of their revolution at will. Is there no meaning in the fact that the most cunning device of human ingenuity for making a point travel, under simple laws, through the greatest variety of paths, should thus prove to be substantially the same with that adopted in the very creation of the human frame, for enabling the hand to guide its tools with freedom and accuracy?

VIII.

GEOMETRICAL INSTINCTS.

Since our fellow citizens of all the animal kingdom are, like ourselves, dwellers in space and time, it is necessary for them, also, to have ideas of distance and direction in space, duration and lapse in time. Ideas gained by sense-perception seem to furnish them, as us, the data for reasoning; but ideas of direct intuition do not appear to afford to them, as to us, objects whereon to reason; but merely serve, as certain of the kind do for us, as the stimulus of desire, and the guide of unreflective action. These intuitive ideas, perceived by inward sense, but not, perhaps, distinctly eliminated in consciousness from co-existent ideas, are, in the lower animals, called instincts; and when used in like manner by us, not as propositions for conscious reasoning, but as the grounds of instantaneous judgment or action, they have the various names of instincts, feelings, promptings, conscience, or genius, according to the nature of the objects to which they relate.

Geometrical instincts are common to us with all the animate races. That instinctive trigonometry, for instance, by which a child, of a few months old, learns to tell the position of any object, to which his two eyes are directed, is probably exercised by all animals with two eyes capable of being turned upon a single object. The most striking instance is popularly believed to be the young quail, which is said to run, as soon as hatched, freely about, pecking at minute objects with as true an aim as its mother's. I have seen a Setter's pup, sired by a Pointer, when a few weeks old, point at a piece of anthracite with all the accuracy of its father, which it had never seen. In 1843, a toad, frequenting the garden at Burgoyne's Headquarters, in Cambridge, and losing by an accident the sight of one eye, was for a long time unable to aim his tongue, with certainty, at the overloaded bees, who, returning, missed the threshold of the hive, under which the toad, expecting such misfortunes of his insect neighbors, was accustomed to sit and await their fall. In time he, like human beings who have had like accidents befall them, learned to substitute optics for trigonometry, and instead of solving triangles, with a base and adjacent angles given, decided on the position of objects requiring a certain focal adjustment and direction of his remaining telescope.

But how complicated the action by which he proved this, if I may use the expression, unconscious knowledge of the position of the bee! A single conscious volition, and his tongue, which is rooted in the front of his mouth, with the tip lying far down within his throat, flies out and back like an electric spark, having taken the bee up on its tip, and thrust her down the throat of the toad. His calculating engine, set by the adjustment of his eyes, not only computes the exact curve in which the tip of his tongue must move, but the exact force and velocity with which it must be sent, in order to accomplish its mechanical errand. The same marvelous unconscious calculation is proved when a boy hits the mark, at which he aims a stone, or when the expert player at billiards strikes his ball on exactly the right part of the ball, in exactly the right direction, and with exactly the right force, in order to make it pursue a long course, partly curved and partly straight, with rebounds from the cushion, and rebounds from other balls, and come to rest at a determined place. He does not know the difficulties of the problem he has solved; he does it with as little of conscious calculation as that with which the toad snaps up the bee; but this only renders the more striking the wonderful perfection of the muscular and nervous organization, as machinery adapted to describe geometrical figures and solve mechanical problems of great perplexity.

The architectural or nest-building instincts of animals show the geometrical and mechanical knowledge of the Creator of animals in a very conspicuous manner. Men invented and used the arch long before human mathematicians solved its theory. Many other of our mechanical inventions, and some of them, as the barrel and the potter's wheel, for example, of a wonderful kind, have an antiquity that long antedates abstract mathematical thought. We, reasoning, discover the principles underlying our inventions, and thus improve science, which again suggests new inventions; so that human art and human science stimulate and foster each other to endless competition and endless progress. The lower races have, apparently, no abstract thoughts, no intuitions, that are brought, with consciousness, among the data of their reasoning; in other words, they appear to have no science, and hence their progress in the arts is so slow as to appear stationary. But their instinctive judgments appear, frequently, more accurate and wonderful than those of men. To see the republican swallow, coming through the air, fold her wings at precisely the right moment, and when at precisely the right speed, in order to enter softly and smoothly her earthen bottle, makes the art of the most skillful coxswain seem rude in comparison. The weaving of the bird's nest is in the case of the African grosbeak carried to a degree of perfection that vies with that of the nicest works of man, unaided by machinery.

But the architectural work of insects is most wonderful, and none more so than the familiar honeycomb. Always admired by men, from the earliest ages, it was, at the beginning of the last century, discovered by Maraldi, of Nice, to embody distinctly the complicated geometrical conception, of forming cells to contain a fluid mass, with the greatest strength, the greatest economy of space, and the greatest economy of material. The paper-making wasps make rude approximations toward the solution of the same problem, but inasmuch as the larger part of their material is cheaply gathered from the surface of wood, there is no call for so strict an economy. The bee, needing a water-proof material, yet finding the resin of trees too adhesive to be worked with facility, confines her use of such resin to the places in which she needs especial strength, or especial resistance to moisture; and, for her ordinary work of cell building, uses a material wholly secreted by the glands of her own body. Her cells are approximately hexagons, which hold more, and have shorter, and therefore stronger, sides than any other figures which could be packed without waste of room. They are set, with economy of room, base to base; and still further strengthened on the bases, by being set one against a part of three, so that the bottom of each cell is supported by three partition walls on the other side. Finally, and most curious, the bottom of each cell is depressed in the centre to about that degree which will save most by diminishing the height of these supporting partitions without increasing too much the area of the floor which rests upon them. I say about that degree; and the accordance of the average cells, in a normal piece of comb, with theoretical perfection as determined by the calculus of Newton, is very close. We should not expect perfection, because the perfection of the artisan is to be measured, not by the perfection of his results, but by the perfection of their adaptation to his end. He were a poor

GEOMETRICAL INSTINCTS. 41

farrier who polished his shoes with the care that a dentist bestows upon his gold filling. Nor would the bee be a wise economist if she wasted time in bringing to theoretical perfection her saving of wax. What the bee's conscious aim is, in the construction of the cell, we may or may not at some time discover. That she has conscious aims is evident, from her adaptation of the form of the comb to circumstances, and from her ingenious contrivances, not only to repair mischief, but to guard against threatened evil. But it is equally evident that in the formation of the bee, and in the inspiration of her instincts, a knowledge and wisdom presided, to which the whole question of maximum and minimum lay open countless ages before human thought grappled with its problems.

It only requires a more intimate acquaintance with the habits of any animal to discover the adaptation of its instincts to its organization. The apparent instances to the contrary arise from want of patience and thoughtfulness in the observer. I stood, one evening, at early dusk, watching the movements of a curious insect on the inside wall of an open shed. Its body, a little over an inch in length, and very thin, seemed, nevertheless, too heavy for its long and delicate legs, which swayed and trembled under the weight, as it slowly stepped along, with long pauses between each step. It walked on four legs, holding the other two, which were shorter and stouter, extended in front. I presently perceived that it was making toward a fly which had settled, apparently to sleep, upon the board within three inches of my insect. I wished to see what its designs were upon the fly, but so slow were its motions, that I was obliged to wait fully twenty minutes before being gratified. As the insect approached the fly, he slowly extended a very long and exceedingly slender antenna, and touched the fly gently, in various parts, as if to ascertain more precisely its position. He then made a detour, and brought himself, at length, exactly in front of the sleeping victim, with his own head nearly over the fly's head, and began very slowly to raise his raptorious legs high above the fly. I was growing tired of his slow and awkward motions, when, in an instant, my feelings were changed to those of the highest admiration for his great engineering skill; the fly was aloft in air, with the beak of the insect thrust into its back calmly imbibing its juices; while the fly's feet could touch nothing, its wings were both dislocated, and firmly pinioned in the captor's raptorious legs; which, coming down suddenly between the wings, had parted them, dislocated them, and pinned them between the wrist-spurs of those legs and their sharp extremities; then, without an instant's pause, lifted the fly from his feet, and impaled him upon the ploiaria's beak.

IX.

MOTION ETERNAL IN DURATION.

The motion of bodies is not observed to be with uniform velocity. We see bodies at rest beginning to move, and bodies in motion coming to rest. Let us consider these cases a little more closely.

We have said that motion implies force, and that force implies will. Force is the energy of will acting upon matter. But how does the will affect matter which is foreign to will, and over which will, we might therefore suppose, would have no control? To this question we answer that the human will never affects the material thing which it determines to move, except through the agency of material agents of whose existence and motions it may be unconscious. Not to speak of the unconsciousness of all earnest labor, the absorption of the mind in its object, take the case in which the mind is seeking to analyze its volitions, and it will be found that we cannot reach, in our analysis, the first effect of the volition upon the physical frame. In the movements of my hand, although I know it is effected by the movement of certain muscles, and that this is effected through the nerves of volition, yet I cannot trace my will behind the command issued to the hand itself. Hence it may be said that the human will moves even the human body through physical agencies, and the manner of its control over these agencies is known only by Him who breathed into us the breath of life.

In like manner is it with the natural motions which we see on every side. All bodies are moved through the agency of other bodies, and we see nowhere a motion which is not dependent upon physical causes, that is, which is not produced by physical agents. Doubtless, He by whose will all things are moved is not restricted from any mode of action, and He can move bodies independently of all law, and without any intervention of means. Nevertheless, such motion would be miraculous, and out of the course of nature. Our will employs, unconsciously, the aid of nerve and muscle; the Supreme Will employs, with wise designs, the intervention of the laws of impulse, attraction, and repulsion.

But when a body at rest receives motion through impulse, it evidently continues the motion of the impelling body. So that, if the impelling body is put to rest by its contact, the motion is not lost, but only transferred. And this is true independently of the size of the two bodies. The earth is not immovable, and the smallest grain of dust that falls upon it strikes with a certain amount of force.

Again: when bodies act upon each other by attraction or repulsion, the force acts upon both bodies. As surely as two vessels, floating on still water, would both move when one attempted to draw or push the other, so surely must each moving thing move those attracted by it, and all that attract it.

This motion, also, exists, whatever be the relative size of the bodies. In the case of motion produced by direct attraction, — for instance, in the fall of bodies toward the earth, — the motion is in simple inverse proportion to the masses. It is, therefore, capable of easy arithmetical calculation. The weight of the earth in milligrams may be nearly represented by the figure 6 followed by thirty cyphers, or six nonillions. Hence the falling of a little insect, weighing six milligrams, would move the earth the nonillionth part as much as the insect fell. That is, the thirtieth place of decimals is capable of representing the motion given to the earth by the fall of the smallest bodies.

There is, then, in nature, no provision for the destruction of motion, but only for its transference. Motion in a body free from external influences is uniform in velocity and direction; it can be retarded, and apparently destroyed, only by external influences, by impulse, or by attraction or repulsion. But these, we have shown, cannot really destroy; they simply transfer the motion to the interfering body. Hence all motion is eternal; it is communicated to an ever-increasing amount of matter; it is, in each successive particle affected, less in quantity, but never becomes nothing, since the sum of the motion in all the particles remains the same.

It is sometimes affirmed, since the demonstration of the theory that heat and light are undulatory motions, that all mechanical motions finally take the form of heat, or of light. The sound of a tuning-fork, placed upon a mass of caoutchouc, is inaudible, and a delicate thermo-galvanic test shows that the temperature of the rubber has been raised. But a still more delicate test, if it were possible to apply one, might show a quality in this heat that would demonstrate it to have arisen from a fork; just as Clairault's nicer calculation showed the comet of 1770 to have been, before it had its short period within the orbit of Jupiter, a wanderer outside the Jovian sphere.

X.

MOTION OMNIPRESENT IN SPACE.

Charles Babbage, in the "Ninth Bridgewater Treatise," has a chapter concerning the permanent impression of our words upon the air, — a chapter which none have ever read without a thrill of mingled admiration and fear; and which closes with an eloquence that were worthy the lips of an orator, though coming from a mathematician's pen.

Would that Babbage had touched, in his fragmentary treatise, upon some of the inferences which may be drawn from the Newtonian law of gravity, — inferences which would probably have been as new to most of his readers as those which he, with so much acuteness, draws from the law of the equality of action and reaction.

The motion of which Babbage speaks, in the chapter to which we refer, is undulatory, communicated by impulse, and requiring time for its transmission; and the startling result of his reasoning comes from the never-dying character of the motion, keeping forever a record of our words in the atmosphere itself, always audible to a finer sense than ours; reserved against the day of account, when, perchance, our own ears may be quickened to hear our own words yet ringing in the air.

But motion is not only enduring through all time, it is simultaneous throughout all space. The apple which falls from the tree is met by the earth: not half way, but at a distance fitly proportioned to their respective masses. The moon follows the movement of the earth with instant obedience, and the sun with prompt humility bends his course to theirs. The sister planets with their moons are moved by sympathy with earth, and the stars and most distant clusters of the universe obey the leading of the sun. Thus throughout all the fields of space, wherever stars or suns are scattered, they move for the falling apple's sake. Nor is the motion slowly taken up. The moon waits for no tardy moving impulse from the earth, but instantly obeys. The speed of light which reaches the sun in a few minutes would be too slow to compare with this. Electricity itself, coursing round the earth a thousand times an hour, can give us no conception of the perfectly simultaneous motions of gravity. There are stars visible to the telescopic eye, whose light has been ages on its swift-winged course before it reached this distant part of space; but they move in instant accordance with the falling fruit.

True it is, that our senses refuse to bear witness to any motion other than the apple's fall, and our fingers tire if we attempt to write the long list of figures, which our Arabic notation requires to express the movement thereby given to the sun. Yet that motion can be proved to exist, and the algebraist's formula can represent its quantity. The position of every particle of matter at every instant of time, past, present, or to come, has been written in one short sentence, which any man can read. And as each man can understand more or less of this formula of motion, according to his ability and his acquaintance with mathematical learning, so we may conceive of intelligent beings, whose faculties are very far short of infinite perfection, who can read in that sentence the motions not only of the sun, but of all bodies which our senses reveal to us. Nay, if the mind of Newton has advanced in power since he entered heaven with a speed at all proportioned to his intellectual growth on earth, perhaps even he could now with great ease assign to every star in the wide universe of God the motion which it received from the fall of that apple which led him to his immortal discoveries.

Every moving thing on the earth, from the least unto the greatest, is accompanied in motion by all the heavenly spheres. The rolling planets influence each other on their path, and each is influenced by the changes on its surface. The starry systems, wheeling round their unknown centre, move in harmony with each other, and bend each other's courses, and each is moved by the planets which accompany it in its mighty dance. Thus does this law of gravitation bind all material bodies in one well-balanced system, wherein not one particle can move but all the uncounted series of worlds and suns must simultaneously move with it.

Thus may every deed on earth be instantly known in the farthest star, whose light, traveling with almost unbounded speed since creation's dawn, has not yet reached our eyes. It only needs in that star a sense quick enough to perceive the motion, infinitely too small for human sense, and an analysis far reaching enough to trace that motion to its cause. The cloud of witnesses that ever encompass this arena of our mortal life may need no near approach to earthly scenes, that they may scan our conduct. As they journey from star to star, and roam through the unlimited glories of creation, they may read, in the motions of the heavens about them, the ever-faithful report of the deeds of men.

This sympathetic movement of the planets, like the mechanical impulse given by our words to the air, is everduring.

The astronomer, from the present motion of the comet, learns all its former path, traces it back on its long round of many years, shows you when and where it was disturbed in its course by planets, and points to you the altered movement which it assumed from the interference of bodies unknown by any other means to human science. He needs only a more subtle analysis, and a wider grasp of mind, to do for the planets and the stars what he has done for the comet. Nay, it were a task easily done by a spirit less than infinite to read in the present motion of any one star the past motions of every star in the universe, and thus of every planet that wheels round those stars, and of every moving thing upon those planets.

Thus considered, how strange a record does the star-gemmed vesture of the night present! There, in the seemingly fixed order of those blazing sapphires, is a living dance, in whose mazy track is written the record of all the motions that ever men or nature made. Had we the skill to read it, we should there find written every deed of kindness, every deed of guilt, together with the fall of the landslide, the play of the fountain, the sporting of the lamb, and the waving of the grass. Nay, when we behold the superhuman powers of calculation, exhibited sometimes by sickly children, long before they reach man's age, may we not believe that men, when hereafter freed from the load of this mortal clay, may be able in the movement of the planets or the sun to read the records of their own past life?

Thou, who hast raised thy hand to do a deed of wickedness, stay thine arm! The universe will be witness of thine act, and bear an everlasting testimony against thee; for every star in the remotest heavens will move when thy hand moves, and all the tearful prayers thy soul can utter will never restore those moving orbs to the path from which thy deed has drawn them.

XI.

THE SPHERE OF HUMAN INFLUENCE.

The conclusions of the last chapter would need but slight modification, should any future observations reveal the fact that the motions of gravity are not absolutely simultaneous. If gravity should prove to be a mere resultant of the undulations of light and heat, we should gain, indeed, a magnificent illustration of the inspired wisdom which begins its account of creation with recording the fiat, "Let there be light;" but we should not lose the spiritual lessons drawn from the fact that the material universe is bound by gravitation into one sensitively-balanced whole, so that each deed of man is felt in the farthest star, and a perpetual record thereof is kept in the movement of the heavenly orbs.

"Every natural fact is a symbol of some spiritual fact." As motion is propagated throughout all space, and endures through all time, so each change in the spirit of each man affects the state of the spiritual universe, and its influence remains through all eternity. As matter by the law of gravity, so spirits by a law of sympathetic attraction are all bound in one harmonious whole, whereof " if one member suffer, all the members suffer with it." Liebig's law, that a moving particle communicates its motion to adjacent particles, was announced in defence of a mistaken theory of trichinosis, but the law itself is true, and universal in physics, in physiology, and in psychology.

The law of attraction holds the same place of primary importance in considering man, as in considering matter. Our great economist makes the Unity of Law a fact of primal importance in the development of his grand and cheering theses. The eagle saint of the Christian Church declares that God is love; and all the highest religions teach that man is made in God's image. He is the sun of infinite magnitude, the origin of all forces, but not moved by any reaction; we are the particles moved by Him both immediately and mediately through each other. Love is the fundamental law; the sympathy between human souls is always greater than the antipathy; even when, through disturbing forces, the sympathy is for a time neutralized, and the antipathy is developed into hatred.

The influence which a man exerts does not cease with the effect that he has upon his most intimate friends; nor does it flow from the power of his word alone, nor from the mere force of his example. Whatever a man does, or thinks or feels, even in solitude, has an effect upon the world. For, in the first place, it affects himself and his own character; and that character must influence, in some manner, those with whom he comes in contact; influence them in proportion to the strength of his power to affect them, and to the weakness of their power to resist him. A cheerful countenance carries a gleam of sunshine into the darkest alley; a sad face throws a shadow over the hearts of those who pass it, even on a crowded thoroughfare; thus every shade of thought and feeling, expressed in the countenance, or in word, or gesture, or action, produces some corresponding change, slight though it may be, in all souls that recognize, however dimly, the expression. And this change transfers itself, in varying proportions, to ever-widening circles. Thus the spirit and tone of the age is the sum of the individual thoughts, and thus also the individual character of each man is to some extent the product of all the preceding ages of the race.

By the manners of a man, or by his speech, we know whether his companions have been Galileans or Athenians. A close observer in the city can tell, of the majority of strangers he may chance to notice, their age, their character, their calling, the place of their residence, and of their nativity, or that of their ancestors. It would only need a nicer observation, a closer insight, a more searching analysis, to detect in a stranger's heart both the original traits of his character and the modifications due to the influence of all with whom he has been associated. It might be a task no more above a Shakespeare's grasp, than the creation of a Hamlet is above the power of an ordinary man, to trace, in a man's present character, the influence of every person and every circumstance that have ever acted upon him to repress or to develop his powers. It would not require an absolutely infinite intellect to trace the effect of any humble act of an honest man, until it had been seen to have blessed millions of his fellow-men ; nor to show the loss or suffering that have flowed to thousands from an evil deed. It may hereafter be possible for some higher intelligence than ours to read the record of my interior life, written in positive or negative characters, upon the soul of some poor man whom I have never seen, but whom I must, nevertheless, have helped or hindered by my every act and word.

Thus the spiritual universe is bound, by the law of love, under its wider enunciation of sympathetic attraction, into one finely-constituted whole, so that not one heart can throb but all hearts must throb with it. " There is joy in the presence of the angels of God over one sinner that repenteth ; " and to every man who falls into sin, we may say, with deeper meaning than that of the prophet, "Hell from beneath is moved for thee, to meet thee at thy coming."

As by the law of gravity the material universe, and by that of love the spiritual world, so by the association of ideas the world of thought is bound into one whole, whereof you cannot find one thought that is not connected, more or less directly, with all. Nothing known, nothing thought, nothing done, nothing felt, fails to leave a clew by which it may be recalled to memory. Each moment's state of consciousness is connected in a train which reaches back to the earliest moments of life, and shall reach on unbroken through eternity, so that it must ever be among the possibilities of memory to recall the thoughts of any instant. And as the rare occurrence of unusual power, developed by accidental excitement, suggests hopes of an indefinite increase of power, when we shall have laid aside this frame, subject to accidents, so the preternatural manifestation of memory, in certain states of health, warns us that this possibility of recalling all things may become an actual reality in the future life.

Then, as the soul surveys the past, with memory presenting its record of every thought and word and deed, and with an eye quickened to see the influence which each has had, she may sit in judgment on her own character; and the word, which shall be the final judge, may speak through the soul itself. Then, also, as she enters the company of cherubim and seraphim, they will need no record of her good or evil deeds other than that written upon herself; from their eyes, as from her own, there shall, when she is present, be no past sin hidden, and no good thought concealed. In the anticipation of standing before such a tribunal, who can fail to find strength to resist the tempter, and encouragement in striving after good?

XII.

MAGNITUDE.

In the minds of some who have read certain of the preceding chapters there has, doubtless, arisen an objection to accepting the results obtained, — the objection that the results are infinitesimal, and therefore non-existent. *De minimis non curat lex;* and that, it may be said, will be the opinion of the court of conscience concerning the "permanent impression of our words upon the air;" concerning the effect of our motions upon the distant stars; concerning the influence of our character upon the tone of the spiritual universe. These influences will be infinitesimal, and to a large part will balance and destroy each other.

In reply to these objections, I will begin by observing that the balancing of two forces is a real effect in nature, not to be for a moment confounded with the non-existence or destruction of the forces. If the earth do not rise to meet this falling rain-drop, it demonstrates that another drop, or its equivalent, is falling on the other side of the globe.

Let me also concede, at once, that, in the form in which the arguments have been put, there is an assumption of certain laws of physics, which, being laws deduced from observation, may be subject to perturbations not yet discovered. Thus Babbage, in the chapter alluded to, assumes that the wave of sound runs frictionless through the air, the heat developed by the compression being absorbed in the expansion. But the experiments of Uriah Boyden have since shown that there is a slight amount of heat developed by the friction of the wave; and this would slowly, but constantly, diminish the force employed in propagating the sound as sound. Again, I have assumed that gravity is a force acting at a distance, and requiring no time whatever for its transmission; but it may possibly be hereafter shown that its speed is not thus absolutely infinite.

With regard, however, to the main objection, that the infinitesimal may be neglected, the objection appears to me not valid, and to arise from the weakness of the human imagination. "Time and space are great only with reference to the faculties of the beings which note them." In space and time themselves there is no natural unit, or scale of magnitude; these are given only by thought, manifesting itself through material phenomena. Hence every scale of magnitude is relative to the mind employing it, and it is only the Unlimited Mind that can be free from the fetters imposed upon thought by the scale employed.

In some of the nebulæ we have examples of the *spira mirabilis* of Bernouilli, drawn on a gigantic scale, so that the part visible to the telescope is many millions of leagues in extent. This spiral may be drawn, on a smaller scale, by tracing upon a circumpolar map the path of a ship keeping steadily upon any one course, not to a cardinal point. Let us imagine this map extended until its radiating meridians stretch out among the stars, and let us trace upon it a *spira mirabilis*, which, beginning at a distance of one billion kilometers from the pole, is running thirty degrees north of east. The length of this spiral will be two billion kilometers, and it will make an innumerable number of revolutions in reaching the pole. Let us, next, imagine that we have come in upon the spiral, until we are but one kilometer from the pole; the length of spiral yet remaining will be two kilometers, and the number of revolutions still to be made around the pole will remain innumerable. Let us again approach until we are within one millimeter from the pole; the remaining length of spiral will be two millimeters; but the number of revolutions yet to be made about the pole will still be innumerable. This little central part of the spiral, all lying within a circle two millimeters in diameter, will be precisely similar in shape to the whole spiral, two billion kilometers in diameter. If it were possible to look at this central part with a lens that should magnify a quadrillion times, it would appear precisely of the same size and shape as the whole spiral. By your approach to the centre, the scale alone would be altered; the central part of the spiral would be a reduced picture of the whole; its linear dimensions would be the fifteenth place of decimals of the dimensions of the original spiral.

That is, it is the nature of the *spira mirabilis*, that whether you move inward toward the pole, or outward away from it, the part between your position and the centre remains always exactly of the same shape, differing only in scale. Let us then approach to within the hundredth of a millimeter from the pole; the central portion becomes now too small to be seen distinctly by the naked eye; but it is of same shape as the whole; and although its total length is only the fiftieth of a millimeter, it still makes an infinite number of convolutions about the pole. On the other hand, if we should run out to the distance of a trillion kilometers, we should, probably, reach the distance of the furthest star visible to the naked eye. That is to say, the unaided eye enjoys a range of vision through about twenty places of decimals in linear dimension. And if we should revise Archimedes' calculation, on the number of sand-grains requisite to bury the universe, we should, consequently, see that sixty places of figures will express the number of grains of the finest silt requisite to fill all space, out to the stars of the sixth magnitude. The sixtieth place of decimals is, therefore, not zero; it is the ratio of the space occupied by a minute grain of fine sand, to the space in which the fixed stars lie. In that minutest space the same forms may lie concealed as those which are illustrated in the whirlpool nebulæ, and which might be conceived as filling vaster spaces. Twenty places of figures, in linear dimension, carry us beyond the limits of sight; but space does not end there; and we might affix figures forever, without arriving at a magnitude so great as to be impossible.

Turning our thoughts again to the minute central portion of the *spira mirabilis*, we could, by the aid of the microscope, see an object whose linear dimensions would stand in the seventh or eighth place of decimals of a meter. In that central speck, visible only to the best microscope, the wonderful spiral would still exist in all its perfection; still making its proper angle with a line to the absolute centre; still being in its fixed proportion of length to the length of that line; and still making an infinite number of convolutions around the pole. And this would hold could we, by the aid of more and more powerful lenses, continually approach the centre until our distance from it stood in the ninth or ninetieth, the nine hundreth or the nine millionth, place of decimals. We may write cyphers, after the decimal point, at the rate of nine million a second, for nine million centuries, but when we finally write a significant figure, that figure is not a cypher; it is significant; and if it signifies the distance which yet remains between our moving point and the pole which it is approaching, then, inconceivably small as the distance is, it has its relations: it is one-half the length of the remaining portion of the spiral; and that portion of the spiral, nearly as it may be without any length or size, still makes its infinite number of convolutions about the pole.

It has been inferred that, because the scale is thus capable of indefinite expansion and contraction, without any destruction of the form, or law of the curve, the contraction might proceed to reduce the spiral into the absolute point at the centre; and that afterward the point might be considered as an absolute nonentity, without destroying the spiral. Thus space and time, it has been said, may be shown to be purely subjective.

But the failure in this attempt to demonstrate the subjective character of space is twofold. A point is not a nonentity; it is a zero of magnitude; yet it has position, or is a position. It is not in the mind, it is in space, fixed in its position, although without magnitude. But although thus real, and indestructible in space, yet being without dimensions or parts, it is incapable of containing a curve except by a figure of speech; by which we either attribute to the point the potentiality of the subjective law; or else speak of a point when we simply mean an infinitesimal space. As we diminish, for example, the scale of the *spira mirabilis*, — by running in upon it towards the centre, and considering only the part yet remaining, which is always similar to the original whole, — it remains real, so long as we have not arrived at the actual centre; but, when that point is reached, the spiral has vanished; there is nothing remaining between us and the pole, for we are at the pole. That pole does not then become subjective; it is a real point; but it does not contain the spiral, except potentially.

In dealing with the infinitesimal and the infinite, the practice of geometers varies; some delight in stating truths in forms which seem false and self-contradictory; others carefully avoid such forms. Both classes are liable to error; the impossibility of clearly imagining the infinite and the infinitesimal is not destroyed by any forms of language ; and the difficulty of reaching true conclusions is not necessarily increased by the use of forms of speech concerning them, literally incorrect. The calculus of Leibnitz is, in a majority of its propositions, literally false ; yet it always leads to true conclusions, in the hands of one who can distinguish between the letter and the spirit.

The best geometer can, however, err ; as, for example, some have said that the point approaching the pole, in a *spira mirabilis*, can never reach the pole, because it will always have an infinite number of revolutions to make before reaching it. The premise is true, but the conclusion false. The point will always, until it reaches the pole, have an infinite number of revolutions to make before reaching it. But if the motion in the spiral be with uniform velocity, then the revolutions will be with increasing velocity ; and finally with infinite velocity ; so that the infinite number of revolutions will be accomplished in a finite time. To take the particular example we have been considering, — the point moving at a constant angle of sixty degrees with the line to the pole, — if the point moves uniformly at the rate of a slow walk, say four kilometers an hour, then it will reach the pole in precisely one hour from the time that its polar distance is two kilometers. But, while the motion of the point is uniform, at four kilometers an hour, and its approach to the pole uniform, at half that rate, the angular velocity constantly increases, being always such that (if it could remain uniform) it would carry the point around the pole in about three and five-eighths the time that remains of the hour. Thus, when at the distance of a meter from the pole, the revolution would carry it around nine times a minute ; while at the distance of a millimeter the revolution would be at the rate of one hundred and fifty circuits a second ; the rotation thus increasing in rapidity, in direct proportion to the approach to the pole. But in less than the one five-hundredth of a second the hour has expired; the centre is reached ; the whole spiral has been passed over, and the point, having made innumerable revolutions, is at the pole. The pole does not contain the spiral ; although it may be said to contain it, in the sense that the spiral, with an infinite number of coils, is contained in any portion of space around the pole, however small, even if smaller than any portion that can be measured, named, or imagined. Thus the point may be said to contain the spiral potentially, so that the spiral would become actual, could the point be magnified.

This illustration of the logarithmic spiral has been chosen and fully expanded, not only because of its peculiar adaptation to the illustration of similarity, and its curious property of always having, even when reduced to an infinitesimal size, an infinite number of coils, but for the historic associations with that curve which Bernouilli would fain have had carved upon his tombstone. The same conclusions would be reached did we consider the shrinking of any other forms in just proportion. It is conceivable that the entire universe might be altered in size, and if proportional changes went on, in every part, in all the forces acting upon it, and in the faculties of all creatures, it would not be in the power of human beings to discover that the change was made. Were the universe thus reduced to the twentieth place of decimals in linear dimensions, and the requisite changes made in the forces of nature and in the faculties of man, the whole stellar system would be contained in the space now occupied by a grain of sand. A second reduction might take place, and a third, and so on forever ; and, unless the rate at which the changes were made increased, the universe would still remain to its inhabitants as large and grand as before. But let the rate increase so as to bring the universe to a point, and its actuality would be gone, and its potentiality alone remain. Still that potentiality would be objective to the Creative Mind; objective in the point. Sweep the point away, and the universe would exist only, as in Erigena's "Second Division of Nature," subjectively in the Divine Thought.

We thus reach sublimer conceptions of the immeasurable gulf between the human and the Divine Mind by holding to the veracity of that intuition which pronounces space and time indestructible objective entities; and renouncing the philosophic wisdom, made popular by Teufelsdreck, which accounts them mere modes of human thought. To the human mind there is no unit of space or of time, save those given in the creation; we cannot even imagine any unit not thus given. To the Divine Mind alone belongs the possibility of deciding on the scale of creation, and deciding what men shall consider large or small, brief or lasting. As the human eye requires the aid of the telescope at one end of its range of vision, and the microscope at the other, so all our faculties are limited to that which is neither too great nor too small for us. But, in regard to the works of nature, we neither discover limits, nor are we compelled by any mental necessity to suppose that there are limits. The bounds of the universe are independent of the weakness and limitations of our powers of imagination.

There is, therefore, no impossibility in the speculations of Lovering, published in the "Cambridge Miscellany" in 1842; that the atoms of our universe may be stars and suns of a smaller one, composed in like manner of infinitely smaller stars and suns; while our constellations and solar systems may constitute only molecules in a vaster world. The infinity of space would almost seem to demand such an arrangement to utilize its wastes. The human mind, fettered by the body, seems in such speculations to show its kindred to the Infinite Spirit, to whom

> "There's nothing great appears,
> . . . There's nothing small."

And these speculations are not confined to a few learned men, whose studies lead them naturally to the theme. Fifteen years before the publication of Lovering's paper, I myself, a child, heard other children supposing that this universe might be the atoms in a crumb let fall by a giant; and that the slow precession of the equinoxes might be the rotation of that crumb, in the air, as it fell to the ground, in an infinitely larger universe.

The true greatness of a work is in the thought which it embodies, not in the scale on which it is wrought. The idea or law of the *spira mirabilis* has a fascinating beauty to a geometer; but to such a mind it is a matter of perfect indifference whether that spiral be illustrated in the minutest shell or in the largest nebula. I have been sometimes as much moved at a slight wood-cut outline of a mountain range as at the sight of the vast masses themselves. The Dead March from Saul will express grief on a grand scale, a sense of human weakness resting in unshaken confidence on the Divine strength, whether played on a single instrument or with a full orchestra.

Our speculations on the scale of magnitude become, therefore, unimportant. Whether there are, or are not, infinitesimal worlds included in the atoms of our worlds, and infinite worlds in which our systems are atoms, the grandest and most inspiring object for our contemplation is the law, plan, or thought, on which the universe, within reach of our faculties, is built. We need not reduce it to a point, or dissolve it in ideal subjectivity, in order to show its unity. The universality of gravity, the co-extensive universality of light, which by the spectroscope has shown that some, at least, of our chemical elements are universally diffused, the correlation of forces, the adaptation of all parts of the universe to each other, all tend to confirm the sublime conclusion that the universe is the expression of one infinitely complex, yet infinitely simple, connected thought, in which all was foreseen, and all comprehended at a single glance by the Intelligence which framed it. The aim of science is to develop and trace the connection of the parts of this intellectual whole; the end of religion is to interpret for the heart and soul the lessons given through this intellectual form.

XIII.

CHANCE AND AVERAGE.

When two phenomena arise from entirely independent causes, the relation of one to the other is said to result from chance. The disposition to consider chance an actually existing cause is so great, that men have, in all ages, personified, and in some nations even deified it.

In the highest contemplation of the universe, as the realization of one grand conception of the Divine Mind, it might be thought that the idea of chance would be excluded, because all phenomena would then be regarded as springing from a single cause; all the minutest events would be considered not only as foreseen, but as intended; as the necessary results of the original thought made actual in the universe.

But the idea of chance, of relations in events springing from independent causes, is so positive in its character, that we are unwilling to concede it to be a mere result of the weakness of the human mind, of our inability to rise to a habitual contemplation of one First Cause. It seems more like a direct gift of power; a power to apprehend some really occurring phenomenon in nature. As such, it forms the basis of a distinct and valuable calculus, applicable to important economic questions of assurance and annuities, and to weighty scientific problems, as a test of hypotheses, and a criterion for rejecting doubtful observations. The successful application of this calculus of probabilities to so many actual problems in the universe is a demonstration that, however difficult it may be to reconcile the conception with our ideas of "foreknowledge absolute" in the single Creative Will, we must, nevertheless, admit into our theory of the world the conception of independent causes, leading to what may be justly called accidental results.

The reconciliation of this contradiction, so far as reconciliation is possible in our finite minds, is probably to be found in the consideration of averages. In our human work we frequently act upon a multitude of individual objects, without special designs in regard to each, but with a general regard to the average action and to the total result. The sower does not consciously choose the position in which a single grain of his wheat shall fall, yet designs and accomplishes an even cast of a given quantity of seed to the acre. The causes which determine the position of each grain are so numerous, and their connection so remote, that they may be considered, for one grain, independent of those for another. In throwing a die repeatedly, in like manner, the causes which determine its position after one throw are so numerous, and so remotely connected with those that determine its next position, that they may be considered independent. Yet the throws are so governed by our will that we may decide, beforehand, how many to make in each minute, — and the positions are so determined by the shape and material of the die, that if it be a homogeneous cube the tendency will be, as the throws are multiplied, to have each side uppermost one-sixth of the time.

This is the law of chance, as applied to averages. And as chance has been personified, and even deified, so average has, by some writers, had divine powers ascribed to it. It has been gravely asserted that the saving of a man from criminal courses only drives another man into crime to keep up the average; as though the present average had an inherent power to perpetuate itself; as though dice could not be loaded without producing a counter-loading in the other player's dice; as though the sower could not vary the size or cast of one handful without immediately varying another handful to keep his field from having more or less seed upon its surface.

The average is a result, not a cause. It is the result of relations that exist between the various causes producing the effects, and may be changed at any time by interfering between those relations; by the dice being loaded, or the sower walking at a different pace; by Jenner's introduction of vaccination, or the discovery of America putting quinine into the physician's hand; or John Howard visiting the prisons, or the Apostle Paul receiving his commission. Great changes thus take place in natural averages; and small changes may be made at any moment by the action of the human will.

It seems not unworthy our highest conceptions of the divine plan to suppose that certain groups of phenomena in nature may, like the sowing of seed by man, be directed and intended for average results, without special design for each individual case. The impassable gulf between the finite and the Infinite Mind would still remain, in the ability of the Divine Providence to select at will any one of the innumerable cases, and employ it as a means to higher and further ends. The winds from the Mediterranean, for example, bring a fixed average of vapor to the summits of the Alps, where it is showered down in countless crystals of snow. These, under slight changes of temperature, contract into minute globes; and these particles of ice, piled up in the mountain basins, press themselves into an almost solid mass, and push themselves, or their earlier companions, down the valleys, grinding off the

CHANCE AND AVERAGE.   67

rocks into powder, which is washed by the melting ice into the rivers and into the sea. The magnitude of these glaciers is limited by the quantity of snow; and, in its turn, limits the quantity of gravel and sand, and the size of the boulders formed by them. To these results, a thousand causes which I have not mentioned conspire. A theist, believing that the glacial system of the modern Alps is part of a divine plan, is not obliged to suppose that this plan includes the position of every individual grain of silt, in the ocean bed, brought from the Alps; is not obliged to make this supposition, even if to his theism he adds faith in a special Providence, and thinks that a grain of sand may be a providential instrument in effecting some great result.

But whatever our explanation of the occurrence of so-called chance among the averages of nature, these chances and averages are frequently adapted to each other with a harmony that seems to admit of no other solution than a reference to the Divine plan which fits each to all and all to each. Enthusiastic students of the calculus of probabilities sometimes represent all human judgments as the result of a calculation of chances; and our certainties are said by them to be merely propositions, the truth of which is infinitely probable. Many of the arguments of natural theology, so called, can be very conveniently put into this form. In the formation of planets around the sun, according to the nebular hypothesis, the chances were small against an order which should fail to preserve the stability of the system; and the present harmony of distances must be referred, directly or indirectly, to presiding thought. In the formation of the solar system, the chances were small that this particular planet should have its elements mingled in precisely that proportion which has resulted in so full a development of life and of human activity; and the arguments of Prof. Cooke's "Religion and Chemistry" derive from this consideration a demonstrative force.

In the course of this successive development of vegetable and animal life upon the earth, there has been, with frequent mutation, a general permanence. Scientific speculation at the present time busies itself with the question whether the permanence has been real, and the changes sudden; or whether the stability is seeming, and the mutations have always been going on with stealthy step. Whichever of these theories proves to be true, that of Plato, or that of Democritus, there is a seeming stability in the present species, which have lasted without sensible change, except the extinction of some kinds, for thousands of years. According to the theory of Democritus, as revived in our days, this arises from the fact that the species at present existing are the fittest for the existing epoch, and thus survive. According to the rival theory this fitness arose from no blind struggle for life, but in accordance with a Divine plan, fulfilled by divine power. In ordinary cases the judgment may possibly remain suspended, whether to suppose the Divine Will acted in reference to the perpetuation of a species by some general law, covering many species, or by special adaptations to one. These cases may therefore be dismissed from the argument. If we grant that a blind evolution by natural variation and survival of the fittest will explain them, it must also be conceded that an intelligent adaptation of the organism to its medium will also explain them. But there are other cases, in which the imagination runs riot in vain for any "sufficient reason" outside of the will and purpose of the Creator, fulfilling an original plan. These cases are like "experimenta crucis," — the theory that fails to hint at a possible explanation fails to explain the universe.

Such, it appears to me, are the cases in which the fecundity of a creature is in inverse proportion to its chances of life. I would by no means say that these are the only points in the animal and vegetable economy which the evolution theories appear to me to be utterly incapable of explaining; but they are the cases which fall under the head of average and chances, and demonstrate that the Eternal Thought which planned this present world comprehended all and more than all which is included in our modern calculus of probabilities.

If the ovaries of the Dodo contained one thousand ova, and if, on an average, less than one of these grew into an adult female Dodo, with equal chances of propagating her kind, it is evident that the Dodo must become extinct. If, on the other hand, two of these ova were impregnated, and came to maturity, it would take but a few generations of the bird to cover the earth, and exclude all other beings. This is prevented, it may be said, by the struggle for life. But the fecundity of each species must be exactly proportioned to the chances of failure in that struggle.

The horse-hair eel is said to lay several millions of eggs; let us say five million. Why this enormous fecundity? Because the chances of the eggs coming to maturity, as eels, is so small. In order to keep the species in existence, two in five millions (if the sexes are of equal numbers) must succeed in escaping all the dangers which beset the eggs and the young in the brook, and then succeed in finding, near the brook, crickets or grasshoppers into which they may penetrate. These grasshoppers must escape their enemies, and survive the depredations of the hair eels, until the latter reach maturity, when they must escape near enough to a brook to find their way there, and meet hair eels of the opposite sex. The chances are two in five millions, let us say, and the creature lays five millions of eggs. Did she average but four millions, the race would in a few years become extinct; did she average six, the creatures would multiply in a few years beyond all bounds. The permanence of the species for so many years demonstrates the accuracy with which its fecundity is proportioned to the slimness of its chance in the falsely-called struggle for life.

XIV.

PHYLLOTAXIS.

The reader may be interested in a more detailed development of the arguments briefly alluded to upon pages 5 and 6, 35 and 36.

There is a rule in arithmetic called the rule of False, or the rule of Position. It is the most general, and the most useful of all rules in the art of computation. Its method renders it applicable to every problem in which the accuracy of an answer can be tested. The rule may be briefly stated as follows: Guess at an answer, and test by numerical computation the accuracy of its results. If the results are accurate, the answer stands the test and was correct. And, if the results are not accurate, the error affords data for estimating the error of the assumed answer; and making a second better guess, to be tested as the first was; and thus to afford data for a third guess, still nearer the truth.

This method of hypothesis and verification is applicable not only to arithmetical problems but to all questions of practical science. It may even be said to be the general rule by which the finite mind approaches every truth which it can approach; and thus reaches every truth which it can reach. In some minds the process is rapid and with a very obscure and evanescent consciousness of its operation; in others slower and with distinct knowledge of the steps; and in all minds the rapidity and ease of the process vary with the nature of the problem. But in regard to all phenomena, which require a cause or theory to account for them, we are not able to draw the theory out of the facts; the theory comes from our own minds; we place it among the facts; and then adopt, modify or reject it, according to its agreement with the facts and ability to explain them. Sometimes the theory explains the facts perfectly, but goes no further; as, for example, when we assume a centre for a circumference passing through three points; the final verification of the centre, by proving it equidistant from the points, gives us nothing further to do; we have found the centre, that is all. But, in more complex cases, it frequently happens that the theory proposed and tested, for the explanation of one set of facts, proves to be also a satisfactory explanation of other facts; its value and authority as a theory are thereby immediately greatly enhanced. Nor is it in the physical sciences alone that this increase of certainty and value may be found; the operations of the intellect are similar, on whatever class of objects it is operating; and theology itself may be lawfully treated, to a certain extent, as Spinoza treated it, by mathematical methods.

Thus Maupertuis assumed it as a fundamental principle that the Divine Being, being unerringly wise, would waste no energy; that everything in nature must therefore be done with perfect economy of force. This theological dogma is called, in mechanics, the principle of the least action; and is, in that science, an invaluable and fruitful principle; as may be readily shown even to those not skilled in mathematical analysis. Take for example the mechanical theory of light. How shall light move in a uniform medium, according to this principle of the least action? Evidently in a straight line, since that is the shortest distance between two points. How shall light be reflected from a polished surface? Evidently so as to make the sum of the incident and reflected rays the shortest possible in going from the point giving light to the point on which the reflected ray shines; and a simple calculation shows that this requires the incident and reflected ray to make the same angle with the reflecting surface. How shall light be refracted in passing from one medium to another? Here, again, we must have the sum of the incident and refracted ray such as to make the whole power expended as small as possible; and a simple calculation proves that this requires the sines of the angles of incidence and reflection to be in proportion of the difficulty of moving in the two media. These instances show the usefulness of the principle of least action in physical inquiry. These results concerning light are amply confirmed by experiment; and shown to give the manner in which a ray of light actually behaves.

But when we turn the argument about, and from this agreement of the experimental results of observations on light, with the principle of the least action, would argue that the creator of light uses a perfect economy of force, and is therefore unerring in wisdom; we feel at once that the argument is not absolutely conclusive. It creates a presumption, but does not force a conviction. We see that the motion of light in a straight line, reflection at an equal angle, and refraction by Snell's law of sines, involve perfect economy of force; and yet these may be necessary results of the constitution of the luminiferous ether, not foreseen when the ether was constituted. In order to make the argument from morphology or from teleology conclusive, the instances of the divine thought and forethought must be such as cannot be explained from mechanical or mathematical necessities. The fact that Maupertuis's theological doctrine of the divine economy of force has led, or could have led, to such numerous discoveries of the actual laws of physics, certainly creates a strong presumption in favor of its theological truth. It shows that Bacon's sneer at barren virgins consecrated to God is wholly uncalled for; here is a religious tenet, a purely theological doctrine actually giving to mathematicians and physicists more knowledge of the universe than they could get from observation of nature, without its aid.

There are various other points in which this theological hypothesis of an infinite wisdom directing the world can be subjected to tests; and in some of them it stands the test, so perfectly as to rise rapidly toward a settled and demonstrated theory; even upon these lower grounds of the understanding, and independent of the intuitions of the higher reason from which Maupertuis and the theologians announce it. Let me briefly touch upon two of them.

In the early history of astronomy it was assumed, that the law of the movement of the heavenly bodies must be a perfect law. The planet, swinging free in space, is subject to no interruption from finite hindrances, but moves under the influence of universal law alone; therefore its motion is perfect. This grand theological conception is worthy a place beside Maupertuis's principle of the least action. But, in attempting to verify this hypothesis, the ancient astronomers made another assumption not so fortunate and trustworthy; the assumption that circular motion is the only perfect motion. They assumed that the orbit of a planet is circular. But on putting this to the test it failed to account for the appearances in the heavens. Still this did not shake their faith in the perfection of circular motion. They only thought that there must be a combination of circles. Instead of a single arm carrying the planet, they put an arm upon the end of an arm; and conceived it as rotating twice, while the first arm was rotating once. If the second arm be very short, in comparison with the first, the path of the planet carried by it would differ but slightly from a circle; while if the arms are nearly equal in length the path would be very different. But they soon found that no imaginable proportion between the lengths of the two arms would enable them to represent well the planetary motions. Still adhering to the circular movement, they place a third arm at the end of the second. This third arm was to revolve three times, while the second revolved twice, and the first arm once. By giving proper proportions to the lengths of these three arms the orbit of any planet could be described with tolerable accuracy. By the addition of fourth and fifth arms, to revolve respectively four and five times, while the first revolved once, the positions of the planets for a single revolution in their orbits could be given with the utmost nicety.

These epicycles of Hipparchus are no longer used in astronomy; but they have been used in modern days, in other departments; in what might be called the statistics of physics, — tabulated observations. It has been shown that with from three to six arms, rotating each with a rapidity proportioned to the number of the arm, beginning at the centre; the end of the outer arm may be made to move in any path required, provided we may fix the length of the arms as we choose, and put them in what position we choose, at the beginning of the motion. And it is justly esteemed a grand work of Hipparchus and of modern analysts (including our own Peirce) that they should have invented and perfected by two thousand years of study so remarkable a result, as that of describing any outline whatever, by the simple device of rotating, simultaneously, three or more radii linked together by their ends. Yet in nature a similar device had been used in a rude form at least from the cozoic ages; and in a perfected form, even better than that of Hipparchus and incomparably more practical and rich in results, from the very advent of man upon the planet. The difference between the epicycles of Hipparchus and those of nature, is that in the former, the ratio of the rapidities of rotation is fixed, and the ratio of the lengths of the rotating arms is varied at pleasure; while in nature it is the proportion of their lengths that is fixed, and the proportionate velocity of rotation is varied at pleasure. For what is this right arm and hand of man but a linked series of radii in which, from the mechanical necessities of the skeleton, the length of the spokes is fixed, but the resulting stiffness is much more than compensated by the variable proportions allowed to the velocities of rotation. The end, free motion in any direction, could of course be attained without a skeleton; as for example the tip of the human tongue can readily be trained to move in any conceivable path within the buccal chamber. But when strength and dignity have been given to the frame by a skeleton, and mobility by articulations, the articulations are exquisitely adapted to the psychic needs of the species. No mechanical necessity can be conceived as producing this adaptation; nor can I see the probability, or even possibility, of this adaptation having been produced by the mutual reaction of the psyche of the animal and its environment.

In the case of man, his immensely varied mental and spiritual powers and capacities; his need of actions incalculably more varied than those needed by the brute; whether we regard his movements as aimed to produce physical or mental effects; require that he should have much greater freedom of movement under definite control. And this is accomplished for him, partly through the perfection of his general form, so "express and admirable;" but more particularly by the epicycloidal movements of the hand. The shoulder is the centre of these movements, but it is not rigidly fixed, it has a proper motion of its own by the sway and torsion of the trunk, by the movement also of the lower limbs, by which the whole body has a capability of motion. But assuming, for the moment, this moveable shoulder-blade to contain in its socket the centre of the epicycloidal movement of the hands; we have first the humerus, swinging freely in all directions, within a cone of about 120°; which is accomplished by what is called a ball and socket joint. The second link consists of two bones in the fore arm, hinged to the first link by a hinge joint;

PHYLLOTAXIS.   77

but this joint receives an equivalent for ball and socket freedom, by the humerus rotating about 90° on its own axis. The third link is the metacarpus, jointed to the arm by a compound hinge, which, however, has an equivalent for ball and socket freedom by the peculiar semi-revolution of the two bones of the forearm about each other. For ordinary movements these three links suffice; the phalanges being kept in a fixed position with reference to the metacarpus; just as three arms suffice for the easier curves in the Hipparchian epicycles. But for the nicest delicacy there are the three joints of the phalanges, perfecting the instrument, just as additional short radii are sometimes needed to perfect the epicycles.

Every one is familiar with the fact that a trained hand can sweep, with a crayon, any outline or figure whatever upon the blackboard. What I am endeavoring to show is, that this ability is furnished by an adaptation of the skeleton and muscular system to freedom of motion, and to freedom under control; by a system similar to the Hipparchian epicycles, famous in astronomy and physics. In fact, every motion of the finger ends, so long as the feet remain in one spot, is an epicycloidal motion; the multitude of links giving it a practical infinity of possible forms. Even the six links from the scapula to the finger-ends give an incalculable variety of possible forms, which our imagination cannot distinguish from an infinite variety. Here then is a wisdom and skill in the adaptation of the body to the mind of man, which seems to me an irrefragable proof of the existence of a wise designer of the animal kingdom.

Let us turn now to the vegetable kingdom for the second instance by which I would test the hypothesis of the world being the work of an infinitely wise Builder. Suppose such a Builder to design a part of the vegetable kingdom to grow by leaves and buds on an ascending axis; the leaves to be the lungs and stomach of the plants, and to require light for the fulfilment of their functions. In the usual growth of such plants it will be manifest that light from the zenith will be most valuable. Side lights will be more apt to be cut off by neighboring plants or other obstacles; and, even if not cut off, may have and will be likely to have the highest pitched and most valuable vibrations absorbed and destroyed or diverted by passing horizontally through so much more of the lower atmosphere. The first necessity of a leaf, therefore, will be zenith light. And it will be expected of unerring wisdom, we might almost say it would be expected of the divine justice, that each leaf should have the fairest possible chance to have zenith light. In other words, it will be expected that the leaves of plants should not be placed one over another; but they should be scattered around the stem in such way as to give each the least shade, and the most light possible. Of course it will not be essential that this should be done with perfect accuracy; practical ends do not need theoretical exactness; on the contrary, it is a mark of a poor workman to have him give a degree of finish and exactness incommensurate with the nature of the work; finishing, for example a kitchen clock as one would finish a clock for an astronomical observatory; that is waste, not economy, of power. But we should expect in plants built by an infinitely wise creator to find distinct evidence that a general plan, for the accurate distribution of the leaves around a vertical stem, was in operation; not distributing them in exact conformity with the plan, but near enough for practical purposes; and evidently showing a perfect knowledge of the perfect plan.

But what would be a perfect plan? Like the old astronomers, we are assuming that the plan of creation is perfect; let us not too hastily assume, as they did, that we know what a perfect plan would be. But, as near as we can see it, a theoretically perfect plan ought to be a symmetrical plan; to go on by a general law;

and to place the leaves always as far apart laterally as possible, as we go up the stem; so that they shall shade each other as little as possible. The first leaf being at the bottom, the second one going up must be nearly opposite, and thus divide the circumference of the stem nearly in halves. Yet when the third leaf comes out still higher, the circumference, as you look down upon the plant, must be nearly divided into thirds. When the fourth leaf is added, the circumference must be nearly divided into quarters; when the fifth leaf comes out, nearly into fifths; and so on. The distance round the stem, the angular distance, as you look straight down on the end of the stem, between any leaf and the one next above, or next below it, must evidently be between ⅓ and ½ a circumference; or else between ½ and ⅔; were it not so, the first three leaves could not fulfil the requirements of the problem.

These requirements with reference to the first three leaves can easily be put in a simple algebraical equation. The second leaf must divide the circumference in halves as nearly as the third leaf makes thirds of it. That is to say: The remainder of the circumference, after two angles have been taken out, must be to the remainder after only one was taken, as that one was to the whole. But by a well-known proposition in proportion this yields at once the proportion that the angle is to the remainder of the circumference, as that remainder is to the whole circumference. In other words the circumference should be divided in extreme and mean ratio; a curious expression familiar in geometry, signifying the division of a thing into two parts, such that the smaller part is to the larger as the larger is to the whole.* Division in extreme and mean ratio was invented by the early geometers, before the Christian era, as a means of inscribing a five-sided figure in a circle; but it was never suspected that it occurred in

---

* Let the arc AB be x; the circumference, 1. Let BC and CD each = AB = x. That the halves AB, and BCA, may be in the same proportion as the thirds AB, BC,

nature until 1849 years after that era. The smaller part is nearly 382 thousandths of the whole; and, if any one wishes to calculate it more nicely, he may extract the square root of 5 as far as he pleases, subtract it from 3, and divide the remainder by 2. It is a peculiar fraction; added to its own square root it produces unity; it is also one-third of the sum of 1 added to its square. But, as the square root of 5 is inexpressible in numbers, this peculiar fraction is also inexpressible. Its presence in nature, were it there, could not be detected by any microscope. All that we could hope to find would be approximations to it. If a plant had such a law of growth that the successive leaves had a tendency to follow each other at equal intervals, each .381966 of a circumference horizontally from its neighbor, we could not measure the angle from centre to centre of the leaves, accurately to any thing like the one millionth; we should be content with finding it about 382 thousandths.

If the perpendicular distance between two successive leaves (called in botany an internode), is large, we can readily trace a helical line around the stem, passing in succession through the foot of every leaf. This may be called the main or principal helix.

and CA, we will put $AC:CB = AB:BCA$; or $\frac{1-2x}{x} = \frac{x}{1-x}$. By proportion, we can add each denominator to its own numerator giving $\frac{1-x}{x} = \frac{1}{1-x}$; and this, by transposing the members, is extreme and mean ratio, $1:1-x = 1-x:x$.

Again, that the quarters AD and AC should be as near equality as the thirds, we put $AD:AC = AC:CB$, or $\frac{3x-1}{1-2x} = \frac{1-2x}{x}$; and again adding the denominators $\frac{x}{1-2x} = \frac{1-x}{x}$; then adding the numerators to the denominators, $\frac{x}{1-x} = \frac{1-x}{1}$ which is extreme and mean ratio as before.

The same result would follow at every step; showing that this length of step would always make the nearest approach possible to equal division into as many parts as there had been points placed, whatever the number of steps.

PHYLLOTAXIS.

And it is evident that we may take the very same set of leaves and pass a helix in an opposite direction through every leaf; only in that case, the angular distance between two leaves will be 618 instead of 382 thousandths of the way round.

Let us now inquire, what further visible and easily observable evidence we should have, if a plant was actually constituted with this law of extreme and mean ratio, as the ideal plan of its distribution of leaves; as seems to be demanded by perfect wisdom and justice, and by symmetry. In case of the internodes being comparatively short, this principal helix will wind round the stem with its threads so close, and the leaves so crowded that it will be difficult for the eye to follow the helix, or discover order in the arrangement of the leaves. Let us imagine the leaves on a piece of stem arranged in this ideal order, and numbered from zero, upward in the order of the stem. Split down the stem on the opposite side to the zero leaf, take off the bark, and spread it flat. If the internodes are short, the numbers over the zero will be arranged as those in figure A are arranged, when you turn the figure cornerwise, so as to have 0 at the bottom, and 55 almost over it, a shade to the right of the vertical. If the internodes are longer, the numbers are better represented by B.

FIGURE A.

15	23	31	39	47	55
10	18	26	34	42	50
5	13	21	29	37	45
0	8	16	24	32	
.	3	11	19		
.	.	6	14		

FIGURE B.

```
   15  23  31  39  47  55
 10  18  26  34  42  50
  5  13  21  29  37
  0   8  16  24  32
    3  11  19  27
     6  14  22
```

In either figure the principal helix, joining consecutive numbers, is transformed into a series of parallel straight lines. But we find also that straight lines from any selected leaf in any direction, join equidistant numbers; and every straight line, when the bark is rolled back into its cylindrical form, will become a helix. Thus the lines in figure A, joining 3 to 5, 6 to 8, and 10, 14 to 16 and 18, etc., would become two secondary helices, steeper than the principal one.

Again, it will be observed that the numbers 0, 3, 6, etc., 5, 8, 11, 14, etc., are in parallel straight lines; and would make still steeper tertiary helices on the stem.

Turning once more in the opposite direction, we find yet steeper parallel lines joining every fifth number; such as 0, 5, 10, etc.; 3, 8, 13, etc.

Still closer to the perpendicular, and in the direction of the original helix, we trace the lines 0, 8, 16 and 5, 13, 21, joining every eighth leaf.

With a still greater compression of the length of the stem, and a diminution of the size of the numbers, lines would become conspicuous, joining the leaves 0, 13, 26, etc.; or the leaves 0, 21, 42; with very great compression and very small numbers, even the lines 0, 34, 68, and 0, 55, 110, almost perpendicular, would become visible and conspicuous.

Observe how curiously these numbers are generated from each other. Starting from zero on B, to go straight to 3 you pass between 1 and 2; but $1 + 2 = 3$. Passing from 0 to 5 we go close in between 2 and 3, and $2 + 3 = 5$. Again the passage from 0 to 8 is between 5 and 3, and $5 + 3 = 8$. The gateway from 0 to 13 is between 8 and 5; but $8 + 5 = 13$.

Observe, also, how readily these auxiliary helices will enable us to number the leaves. Suppose this piece of bark, with its footprints of the leaves on it, is before us, but the leaves not numbered. Assume any one as zero. Now here are three parallel helices taking in every leaf. Then upon the one passing through 0 you can certainly write the numbers 0, 3, 6, etc. Here are five other helices crossing the first and including every leaf. Then upon the one passing through 0 you can write 0, 5, 10, 15, etc. You can now pass to any leaf whatever by adding or subtracting 5 or 3, as you move up or down on either one of either set of helices.

Were the stem crowded with a greater multitude of leaves, so that the more perpendicular and more numerous helices were most prominent, the same thing could be done. For example, were there 21 helices in one direction and 34 in the other, and you wished to find the 8th leaf from one which you had chosen as zero, you have simply to ascend two steps, on one of 21 leaf lines, and descend one step on the 34 leaf line; since $42 - 34 = 8$.

It will at once be seen that if the stem of a plant be tough, and of even texture, so that it will bear twisting evenly; then, if the leaves were arranged in this fashion, a twist in the stem would straighten up every alternate set of helices, and flatten down the others. A very gentle twist might, for example, bring the 34th leaf over the zero. Every leaf, the twist being uniform, would then be exactly over the 34th below it. The 34 leaf helices would be vertical lines, and the leaves would be in 34 perpendicular rows. A little harder twist would bring the 13th leaf over the zero; and the leaves would then stand in 13 vertical rows. Twist still harder; the 5th leaf comes over the zero; the five helices become five vertical rows. If your stem is tough enough, twist violently enough to bring 2 over the 0, and your leaves are in two rows, alternately opposite.

Let us now imagine our gentle twist at the beginning to be in the opposite direction, and bring the 21st leaf over the zero; this would give us 21 perpendicular rows; a little more twist in that direction would develop 8 rows; a hard twist reduce them to three.

This supposed twisting of the stem would increase or diminish the theoretically perfect angle, .381966 of the circumference, until, in the case of the two rows, the angle was .5; or, in the case of the three rows, it was .333333. Between these extremes, ½ and ⅓ of the circumference, lie all those arithmetical approximations to the perfect angle, which we have developed by twisting; 1:2, 1:3, 2:5, 3:8, 5:13, 8:21, 13:34, 21:55, 34:89, 55:144, 89:233, 144:377. These approximate angles are formed each from the two preceding, by simply adding the two numerators, and adding the two denominators.

One half is larger than .381966, and ⅓ is smaller; and throughout the series they are alternately larger and smaller than the perfect angle. No other vulgar fraction stands in the line of direct advance. Take for example 7:19, it is not as near as 3:8, nor is 7:20, nor 5:14, nor any fraction with a denominator under 21, except 5:13. This series settles toward the extreme and mean ratio like a vibrating needle settling to the magnetic meridian; the last one given above differs from the true angle by only .001 of 1°.

Suppose we should now take the case of an actual stem crowded with leaves, and should find that an actual twist to the left or to the right would bring the leaves into 3, 5, or 8 rows, what would it prove? From the diagrams A and B it would prove that the leaves were actually arranged at equal angles around the stem, and that the angle was between 1 : 3 and 2 : 5 of a circumference. Suppose, however, that a slighter twist would bring the leaves into 13 or 21 rows; or suppose that the 34 and 55 leaf helices were already conspicuous, without any twist; and suppose that by counting up by 34s and 55s we should find the 377th leaf was, more exactly than any other, vertical over the zero leaf. This would prove the leaves to be actually arranged not only evenly around the stem, but at an angle almost precisely that of the theory.

These suppositions are actually verified. Taking up, for example, a stalk of broad-leaved plantain, crowded with flower buds, or with seed pods; you can, by twisting in one direction, bring out 8 or 3 rows; by twisting in the other, 5 rows, and with a hard twist, 2. Taking up any pine or spruce cone, and numbering a few scales by first counting the conspicuous parallel helices, in each of the two conspicuous intersecting sets, you will find the 13th, 21st, 34th, or 55th leaf come most directly over the leaf taken as your zero.

When the cylindrical stem is contracted into a strobile, cyme or rosette, or into the nearly flat head of a composite flower, the helices are transformed into other curves. Were the stem transparent, and the helices drawn in perspective upon a plane at right angles to it, the eye being the axis of the stem, the helices would become hyperbolic spirals. But the spirals of sunflower seeds make no approach to that form. Were a short portion of the stem, the height being equal to the radius, folded in toward the centre, the distance of the helix from the circumference being unchanged, the helix would become a spiral of Conon, or Archimedes. But the sunflower does not conform to that law.

The helices each make a constant angle with the meridian of the cylindrical stem; and as far as I have yet observed, this is true of the transformed curves; so that on a globular strobile the helix becomes a rhumb line; and on the sunflower head, a spira mirabilis of Bernouilli. In the rosettes of young Œnotheras, Capsellas, etc., the spiral is developed by a different process; and may possibly be a different spiral, but the sunflower certainly approximates (according to my measurement) very closely to a logarithmic spiral; the spira mirabilis.

In the heads of composite flowers these spirals are beautifully conspicuous, and afford an easy method of determining the degree of approximation. The vertical lines have here become radii. There will always be two spirals (one running in each direction), plainly to be seen, crossing at each seed. The value of each of these can be determined by counting round the stem and seeing how many similar spirals there are in each direction. Thus, if there are 8 spirals, as it were, parallel to each other, running round to the right, there they must be the 8 leaf helices; and the 5 leaf helices will be found running to the left, 5 in number. Assuming now any seed as a zero leaf, the adjacent seeds running up the spiral to the right will be the 8th, 16th, 24th, etc.; and running to the left the 5th, 10th, 15th, etc. By running to the left and right alternately you can thus determine the number of any seed as you please. By determining as nearly as possible the centre of the head, you can draw a radius, from your assumed zero and inward. If that radius, after leaving the zero, strikes first on the centre of number 34, the angle is 13 : 34; but if it steers between 34 and 55, grazes 89, and strikes a seed centrally first at 144, then the angle must be 55 : 144.

As I was writing this, I picked up three heads of dandelion in seed, blew off the seeds, and counted the pits in the receptacle. One of them was on the 34 : 89 arrangement; the other two each on the 55 : 144 arrangement. The largest of these was half an inch across; and the error in the position of any two consecutive seeds in the outer row was, therefore, less than one 30,000th of an inch. One of the heads gave an angle a trifle too large; the other two, an angle a little too small; the average was almost exact.

Subsequently, I counted the first ox-eye daisy, and the first sunflower which I saw. They were equally exact as the dandelion, — the sunflower even nearer. It was upon a 144 : 377 approximation; that is, the angle, instead of being 137°.508, was 137°.507, only about a thousandth of a degree too small. The head was about 20 centimetres or 8 inches in diameter; so that two consecutive seeds near the circumference would have an arc of 9½ inches or 24 centimetres between; and this arc was actually about one 500th of a millimetre, less than the ten thousandth of an inch, too small. It would surely be unreasonable to ask for any closer conformity of observation to theory.

Nor is it easy to imagine any cause which necessitates the arrangement. It has been shown that if a cell generate cells on a horizontal plane at equal intervals of time; and each cell begin to generate in the same manner, as soon as it is two intervals old: and the generation be always on a plane, and at the right hand side; then the cells will be arranged like the seed pits in the dandelion receptacle. But this goes very little way,—nay, I do not see that it even starts on the road, —toward showing us the genesis of the ascending helix.

Again it has been suggested that as leaves grow by light and air, they will naturally grow where they have the best chance at getting them; and this in the course of generations would lead them to come out exactly at the right spot. Unfortunately for that explanation, the leaves of plants which need the light and air, and for whose benefit we have invented the law, do not conform to it in such wise as to make it a physical benefit. They grow almost universally on the lowest approximations, 2, 3, and 5 ranked.

Practically, the physical benefit is not felt. And in the part where the highest approximations are reached, in the heads, cones, cymes, etc., the physical benefit is lost through the crowding. We have evidently over-estimated, at the beginning of our speculation, the merely physical necessity of the arrangement.

Of course we can set no bounds to the discovery of physical causes for physical effects; and it is therefore possible that the botanist may, at some day, discover the physical agencies by which this physical arrangement of leaves is effected. But when he has done so, he will not have in the least shaken the theological inferences. The preponderance of the ruder approximations, $1:2$, $1:3$, $2:5$, and $3:8$, ($1:3$ and $3:8$ giving too small, and $1:2$ and $2:5$ too large, an angle) shows that the perfect phyllotactic law is not of practical importance in the growth of plants; they live and flourish on the rudest approaches to it. But the tracing of these approximations up, in such very numerous instances, to the highest degree of accuracy, such as $55:44$ and $34:89$, one above, the other below the perfect, shows that the law of extreme and mean ratio is actually incorporated into the vegetable kingdom. The builder of the plant knew that law untold ages before the geometer invented it, to inscribe a pentagon. These successive approximations point out more clearly and strikingly than absolute conformity to it could have done.

And as its efficient cause thus lay in the divine wisdom and divine power, so its final cause lies also in the spiritual realm. We have come upon it by an assumption that the leaves are treated justly; that each is given the best possible chance at light and air. But while we have learned from our examination of it that the divine Architect knew this need of the leaf, and in providing for it took this absolutely perfect law; we learn also that he knew that perfect conformity was not physically needed; and he therefore allowed these continued and great variations by which the law is suggested rather than thrust upon us; he made the symmetry of the plant potential, rather than actual, and this suggestion, rather than actualization, of the perfect, makes the plant a more valuable teacher and companion of man. The suggestion of infinite perfection, that is beauty.

> The Lethe of nature
> Cannot trance him again,
> Whose soul sees the perfect,
> Which his eyes seek in vain.

The outward eye cannot directly see the division of the circumference in extreme and mean ratio; half the leaves are hidden, and even if we see two consecutive leaves, we cannot tell the precise angle. But the secondary helices of approximation are constantly visible, and give a great geometric fascination to the fructification, sometimes even to the foliage; while in the larger growth of the plant the law secures general symmetry; the variation and concealment through various causes, prevent monotony and give an endless charm of variety.

XV.

NUMBER AND PROPORTION.

It is only at a comparatively late period, in the development of the human mind, that number comes into view, as a distinct object of thought. The idea of number is evolved from things of imperfect unity; as an abstraction from things concrete, tangible and audible. The two hands, the ten fingers, the mother, the nurse, the window-panes, suggest the idea; and it is slowly brought into the field of distinct intellection, by more or less laborious effort. A child usually attains the age of three or four years, before it gives evidence of attaching clear ideas to the names of numbers. Not until adult life do men usually perceive that persons are examples of the most complete and absolute unity.

But this slowness with which the idea of number rises to the surface of consciousness, only shows how very deeply it is imbedded in the soul. At the beginning of conscious life our attention is fixed upon the individual objects presented in sensation. The child at that period,

Nescio quid meditans nugarum : totus in illis :

thinks only of the direct lessons of the outward world. The abstraction of number, and the invisible realities of space, of time, and of the spiritual world escape his attention, until he arrives at a mature condition. But the mature mind perceives that this historical order of the development of ideas is almost invariably precisely the reverse of the logical order of their dependence. For example, space and time seem at first to be abstractions from the observed facts of matter and motion; and it is hastily assumed that the experience of the outward world comes, at first, independently of any perception of space and time; and that these ideas are derived from that experience. This may be true in the chronological, or historical, succession of our distinct analytical attention to the ideas; the actual sequence of distinct conscious attention is, that we first perceive motion, or rather matter in motion; and this leads us to the consideration of space and time.

Subsequent thought will, however, always show that, in the very first perception of matter in motion, we quietly take for granted the existence of space, occupied by matter; and of time taken up in the motion. The ideas of matter, motion, space and time, actually enter the field of consciousness at the same instant; that is, the historical order only relates to the sequence of the acts of attention, by which we separate the ideas from each other. In strict logical connection, the conception of space and time must precede the conceptions of matter and motion; since space and time are the conditions on which motion is alone possible.

In like manner it will be found that although number is a late object of conscious attention; and can be developed as an object of distinct consecutive thought only in a mind of some maturity; it nevertheless stands logically antecedent to every act of intellection. The conscious subject is conscious of an object; and these with the act of consciousness form a tri-unity at the very beginning of any conscious life. If we dare venture so high a flight of thought, we may even say that as the creative Mind must be posited as a logical antecedent to creation; so even in that infinite Mind, considered even as its own object, that same tri-unity existed antecedent to any creation. Of course we speak of logical, and not of chronological antecedence. Whether the latter ever really existed; whether there was creation, in the widest sense of the word, is a matter beyond our most daring flight. But in the first act of intellection, the conception of subject and object implies the conception of number. As with the idea of self so with the idea of space and time; they also necessarily imply number the moment that they are made objects of thought. Even infinite space, although absolutely homogeneous and without distinction of parts, except as such parts are created by thought, has its two elements of distance and direction; and the element of direction has a manifoldness, which by no artifice of ingenuity can be reduced to less than three dimensions. Some modern geometers, assuming that the conception of space is derived from, as well as suggested by experience, have speculated upon the possibility that, in those parts of the universe which are beyond the range of our experience, space may have other properties; more dimensions than three, or a curvature by which two straight lines might include a surface; and so on. Pursuing this speculation they have investigated certain algebraical sentences, expressing these impossible conceptions; and have found the language capable of self-consistent interpretation. They have urged this possibility of self-consistent interpretation as an evidence that the conceptions themselves may be realized in the infinite distance. In spite of this ingenuity Reason sturdily maintains that the properties of space, as we see it here, are invariable throughout the whole extent of absolutely boundless distances; and although we may technically express, in algebraical language, the conception of a circle having unequal diameters, and deduce logically self-consistent results from the conception, yet we can neither make any picture of such a circle in our imagination, nor believe that such a circle can exist in any regions beyond the telescope.

Number, inhering in the primal act of consciousness, follows every step of thought. All intellection, all thinking is the perception or creation of differences and distinctions, unities and resemblances. The definition of chemistry, that it is the identification of the one in the many, and the detection of the many in the one, may be considered also as a definition of all science and of all thought. Number is more prominent in Chemistry, just as Space is more prominent in Mechanics, and Time in Biology; but Number, Space and Time are all three involved in every finite act of intellection. All language bears witness to this presence of the three ideas in every thought; take any word of any language and analyze it carefully, trace back its history and you find in it some more or less apparent reference to number and motion, taken perhaps as typical of spiritual things.

But number, although thus involved in every act of consciousness, even the primal, is not the highest genus; it is a species of relation. The highest unity is the person; and the highest Person, although we speak of Him as Absolute and Unconditioned, stands logically related to His own attributes and to His own creation. Every act of finite intellection involves not only the perception of number, but of other relations also. There is more in the consciousness of subject, object, and relation between them, than the mere perception of tri-unity in the act. There is the perception of more than the numbers three, two, and one. Of course in the more complex acts of intellect, it is still more emphatically true, that the perception of numerical relation does not constitute the whole contents of consciousness. Number is, however, the relation by which the relation of quantity becomes amenable to thought and calculation.

At first, number is generated by the distinction of things discrete, and perhaps different in kind. It immediately becomes in itself an object of thought; and its distinction of many or few becomes the most firmly grasped and clearly comprehended of all relations of quantity; which is at once applied to the measurement of all things that may naturally be counted. But there are many kinds of quantity which are, absolutely or relatively, continuous; and the measurement of the greater or less in these kinds is accomplished by the analogy of the greater or less quantity to large or small numbers. So indispensable to all clear intellection is this relation of numbers to each other, that the Greeks called it λογος, that is word or wisdom; and the Latins called it ratio, or reason; and this is its technical name among mathematicians to the present hour. The ratio of two numbers is their relation of magnitude, not as estimated by the excess of one over the other; but estimated by how many times one is larger than the other. Thus the excess of 6 over 2 is the same as that of 8 over 4; but the ratio of 6 to 2, is 3 to 1; and that of 8 to 4 is but 2 to 1. When we seek to find the ratio of one continuous quantity to another of the same kind, we simply seek to find two numbers having the same ratio as the two quantities. One quantity is considered as the unit, to which to refer the other. Usually as a preliminary step, both are first referred to some artificial unit; such as an inch, a meter, a quart, a liter, a degree of the thermometer, a degree of angle, an hour, a dollar, etc., etc.

Yet many of the most interesting ratios are found not to be equal to the ratio of any two numbers whatever. For example, the diagonal and side of a square, although nearly in the proportion of 17 to 12 are not exactly in the ratio of any two numbers whatever. And in general we may say that the mean proportional between 1 and another number, will very seldom be in a ratio expressible in numbers. Two is a mean proportional between 1 and 4; because 1 is to 2, as 2 is to 4; 3 is a mean proportional between 1 and 9; and so on. This mean proportional, or geometric mean, may be illustrated by letting fall a perpendicular from any point in a semi-circumference upon the diameter; the length of the perpendicular is then a mean proportional between the two parts into which it divides the diameter.

In the attempt to measure quantities, it is frequently necessary to assume a starting point, and measure in both directions; — one is then able in subtracting to obtain quantities less than nothing; like south latitude, east longitude, temperature below zero, deficit in a treasury; such quantities are called negative quantities; they lead us at once to the useful fiction of negative numbers. But this fiction requires the additional fiction of negative ratios. For example, the ratio of 20 above zero to 10 below zero, is not simply that of 2 to 1, which would only give 10 above. It must be expressed by saying it is negative 2; — meaning that it is twice as far from the zero mark, but in the other direction.

But out of this comes a very remarkable case of impossibility; namely the impossibility of finding the geometric mean between a positive and a negative number. The geometric mean between 1 and 2 cannot be expressed in numbers; but the fraction $1\frac{2}{5}$ is nearly it; and no error can be named so small that a fraction cannot be named differing from the geometric mean less than that error. But the geometric mean between 1 above and 1 below zero cannot be expressed by any degree of the thermometer or by any conceivable numbers. It must be 1, but it can be neither positive 1 nor negative 1; since 1 is not to positive 1, as positive 1 is to negative 1; neither is 1 to negative 1, as negative 1 is to negative 1. Certainly zero is not the geometric mean; for zero is not as many times negative 1, as 1 is times zero. What then is this temperature neither above, below, nor at zero? this latitude neither north nor south of the equator? It is called in mathematics (*lucus a non lucendo*) the imaginary; because it is unimaginable. The mathematicians have various symbols and various names for it, and use it freely in their calculations. They have endeavored, with great success, to reduce all quantitative and geometric impossibilities to this one; that is to say, they will give you a correct answer to any absurd question you may devise, provided you allow them to introduce into their answers a symbol standing for the unimaginable mean proportional between positive and negative unity. In geometrical questions this symbol may often be interpreted as signifying the rotation of a line about some point, but no general interpretation has been discovered. It is, however, a singular tribute to the ingenuity of the mathematician, that he has reduced all absurdities, all impossibilities, to this one, — the finding of an operation upon a quantity which shall neither increase it, diminish it, nor leave it unchanged. Give him a symbol, say i, for that operation, and he is as competent to deal with the impossible, as with the possible. But the advantage and power, gained by the use of i, extends also into the realm of the possible and of the actual; and enables the mathematician to assist in delicate and complicated researches of modern physical science, otherwise beyond the reach of man.

If we multiply 1 by 1.00000001, one hundred million times, the product is 2.7182818. It is the second of three famous quantities, pertaining to number alone, yet incapable of exact expression in number; since exact expression would require the decimal part of the constant multiplier to be infinitely small, and the number of multiplications to be infinitely great.

This peculiar number is known in mathematics by the name of the base; (THE base); and may be symbolized by b. It may be defined as the product of 100,000,000 factors, each equal to 1.00,000,001; but is susceptible of various other definitions. Like i, it furnishes the key to many physical problems, otherwise insoluble; and it is very difficult to express, in terms not technical, any reason why it should exert this beneficent power.

The third of these famous quantities is the ratio of the circumference to its diameter; which is nearly 22 to 7, still more nearly 355 to 113; but which cannot be expressed exactly either in numbers, or in mean proportionals between numbers; it is a peculiar, unique ratio; it may be symbolized, if you please, by c; and its immense value in calculation arises partly from the fact that the difficulty of measuring not only circles, but every conceivable kind of curved lines, surfaces and solids, is generally reducible to the calculations of b and c. Add i and you have the means of calculating the inconceivable also.

These three famous ratios, so entirely different in their origin, and so utterly incapable of exact expression in number, may be connected by a very simple bond. If we multiply 1 by 2 ten times we obtain 1024; and if we multiply 1 by 1024 we obtain the same. Since ten equal multiplications by 2 are equal to one multiplication by 1024; we may say that multiplying by 2 is multiplying by 1024 one-tenth of a time; multiplying by 8 is multiplying by 1024 three-tenths of a time, and so on. Using the like expression we may say that b multiplied by itself half c times gives 4.8148. Divide 1 by this number and we get .20788. Next we take the impossible, inconceivable, inexpressible i, and multiply it by itself i times; and can readily demonstrate that it produces identically that same .20788. Here then are the three famous and inexpressible ratios of numbers, — base, circumference, and the imaginary, — bound together in this simple law; to wit: Multiply base by itself, half circumference times; then multiply the product by imaginary, imaginary times; and the final product is unity.

The object of this long discussion of such abstruse numerical relations is to set in a more striking light the marvellous power of the human body as an unconscious calculating engine, alluded to in more than one of the preceding chapters. Number, the creation of conscious spirit, created in the act of consciousness, cannot, in its utmost reaches of abstruseness, go beyond the power of the Infinite Mind. But in our finite thoughts we reach conclusions, like the above, binding together what is conceivable, but inexplicable, with what is neither conceivable nor expressible, in one inexplicable bond.

Yet feeling, or emotion, is a higher, deeper state of consciousness than thought; and often carries us into regions of the Divine thought where finite thought is unable to follow. In some instances the reality of this flight into higher regions has afterward been verified, by the slower laborious ascent of the finite thought, in paths over which feeling had flown. Thus early thinkers declared that in the perception of musical harmony, the soul unknown to herself calculated secretly the numerical ratios of the undulations of the air. Modern musical statics has proved that these secret, unconscious numerical calculations by the musical ear, by the mere aesthetic feelings, have been far more numerous, complicated and precise than the older thinkers had supposed; and that elaborate comparison, calculation and experiment demonstrate the course and progress of music, from barbaric times to the present, to have been in perfect agreement in all respects with laws of number, not revealed until very recently. It is certainly difficult to imagine modes by which experience and habit could have led to this progress, had not the human body been in its very creation formed in exquisite adaptation to those laws.

I have, in another chapter, alluded to Hay's law that beauty of geometric form depends upon the division of the right angle in harmonic ratio. Burke argues with a great deal of misplaced ingenuity to prove that proportion has nothing to do with the cause of beauty; but Hay simply takes the Parthenon, the Temple of Theseus, the Lincoln Cathedral, and other universally recognized types of beauty, and shows that their actual and potential angles do always stand in simple harmonic ratio to the right angle. Here, therefore, is a second instance in which the secret unconscious calculations of the beholders have always recognized by feeling what the labors of Hay have shown to be numerical proportions in architecture.

But I think I have proved that Hay's law is incomplete; and that the eye for beauty recognizes sometimes a far more delicate and involved proportion than that which he assumes. He illustrates proportion of geometric stationary angle by the proportion of vibrations in time, producing a musical concord; and draws his law from an examination of architectural and plastic art. But the beauty of the vegetable world led me to consider that the phyllotactic ratio fully developed in the preceding chapter might also give beauty in artificial forms. The approximations $\frac{1}{2}$, $\frac{1}{3}$, $\frac{2}{5}$, and $\frac{3}{8}$, would be also included in Hay's series. To test this question I have tried various experiments with satisfactory results. A single one will show their nature. I drew two semi-ellipses; in one, the angle (from the end to the middle, and to the mid side) was $\frac{3}{8}$; so that either by phyllotaxy or by harmony it should be beautiful. The other was drawn exactly on the perfect phyllotactic angle; which, as harmony, would be rudely discordant. Showing these two ellipses, without explanation, to many persons, in private and in schools, and taking a vote on the merits of their shape, the vote was unanimous on their *both* being good, very many preferred the phyllotactic angle. They were so nearly alike in their proportions that some observers mistook, in attempted conscious calculation, and thought the higher one flatter, even although joining in the judgment that it was more beautiful.

How much more wonderful, if possible, is that unconscious calculation of numerical relations not yet confirmed by figures, but proved by experiment to be correctly performed in higher departments of art. The harmony of colors unquestionably depends on numerical relations of wave lengths. The workman at a Bessemer steel factory knows when to draw the charge by a secret unconscious calculation of the rapidity of vibrations, millions in a second, taking place in the flame. All the world have by similar, but more delicate computations, confirmed and endorsed Titian's judgment in the harmony of colors. What is a great painting but the conveyance of great thoughts and feelings from the painter to the beholder through medium of numerical ratios of angles in the drawing and of vibrations in the coloring. It is ratio; it is the logos, the incarnate word of the painter, imitating the Almighty Logos by which the heavens and the earth were made, and by which the painter's hand had received the skill to make, the beholder's eye the skill to interpret, the imitation.

The same questions may be asked, the same answer must be given, concerning the higher ends of music. I have demonstrated by hundreds of carefully conducted experiments upon hundreds of persons, sometimes upon large classes in schools, that fully three-fourths of an audience receive from a musical composition the very same moral mood or tone of feeling which filled the composer at the time of composing it. The closing chorus in Beethoven's "Mount of Olives," played upon an organ, without words, gives to a hearer who has not heard it before, and who is not informed of what it is, immediately and irresistibly an impression of solemn awe, of inexpressible majesty, of penitence, yet of the peace of forgiveness; of gratitude for forgiveness, and a sense of reconciliation through mediation. All this is accomplished simply through rhythmic modulations; but not, as in spoken words, through any conventional or artificial associations with them. It is the direct natural language in which Beethoven utters his profoundest, deepest faith. If the hearer have known anything concerning the tenets of the Christian faith, he will be certain that the composer was a Christian believer, uttering through those chords a thanksgiving, in the name of all worlds, for the reconciliation of the world through the cross. What explanation can be given of this high power of the artist, in whatever department, to pierce direct to the heart without the aid of conscious intellection in the head? I see none, except to admit that given in Emerson's " Problem," namely, to admit that the forces of nature, partially under our conscious guidance, are wholly under the control of that same infinite wisdom and love which inspires the artist, and interprets his work to his audience.

XVI.

THE DEVELOPMENT OF FORMS.

It is sometimes said that classification always proceeds by a process of dismissing from attention that which is peculiar to the individuals; and fastening the mind upon some arbitrarily chosen points of resemblance. Examples of this method are found in Linné's artificial system of botany, and in Agassiz's proposed classification of fish by their scales. All classification does not, however, proceed upon this method. Again, it is said that the classification of the organic kingdoms is an arrangement according to the degree of development; that development has been a continuous process; and the degrees have been accidentally, or else arbitrarily marked out. A little examination will show that this view, also, is erroneous.

The classification of the organic kingdoms is unquestionably based upon form, in the geometric sense of the word. An accurate drawing, an accurate plaster-cast, the impression left in a rock, — these are sufficient data for a naturalist to decide the identity of a species. If, therefore, classification is built upon development, it is upon the development of forms in space. All these forms have a certain amount of symmetry; their outlines and surfaces correspond, more or less perfectly, to the geometric law. We have already shown that this conformity to law is obedience to thought. But not only does the form of each individual conform to law; the series of forms, also, as Agassiz has shown, indicates a law pervading the series; and this fact is inconsistent with the opinion that evolution, if there has been evolution, has proceeded continuously, by insensible gradations.

All scientific men, at the present day, admit the reign of law in organic nature; even those who believe in continuous development. The origin of the universal, invariable law of inorganic nature cannot, however, be discovered by an investigation confined to nature herself. That question lies outside the realm of science. The scientific man may consider it, and may answer it, wisely or unwisely; but in thus doing he has left the domain of science, and entered that of philosophy. Science infers the universality and invariability of natural law, by a "simple enumeration" of the increasing number of observed phenomena conforming to law; philosophy takes a firmer ground. She conceives the universality of law to flow as a necessary consequence from the grandest and most certain conclusion of human thought; the being of an infinitely wise Creator. The infinite Wisdom foresaw from eternity the best possible modes of action, and adopted them. No occasion can arise to make Infinite Wisdom change its plan. "By the determinate counsel of the Lord," says the wise Hebrew, "were his works from the beginning;" "He gave eternal order to his works, and to the atoms thereof, for all generations;" "nor to eternity shall they disobey His word." No scientific writer of the nineteenth century can more distinctly affirm the universality and invariability of law than this religious teacher more than twenty centuries ago. Historically, as well as philosophically, the doctrine is the direct outcome of theistic faith. To one whose mind has grasped the conception of the existence of an infinite Deity, of infinite wisdom and power, the whole universe becomes the expression of a single Divine Thought, almost infinitely full of detail, and thus giving endless occupation to our finite intellect; but also possessing perfect unity, and thus flooding us with the fulness of its beauty. In this theological form the doctrine of the correlation of forces, and interdependence of all sciences, was familiar to metaphysical and theological writers, long before its tardy confirmation by the induction of physicists. The natural sciences, like the mathematics, have their postulates; among them is that of the invariability of law; they need also the postulate of the universality of law. The doctrine of a continuous development will be found difficult to reconcile with either.

This doctrine is a virtual denial of the existence of law in a department in which the whole history of science would lead us most surely to expect the discovery of law. Evolution itself has strong antecedent probabilities in its favor; it is only against the mode of evolution by continuous change and accidental arrest, at certain accidentally determined stages, that the present objections lie. The whole course of scientific study up to the middle of this century had revealed more and more clearly the presence of law, governing the atoms, and molecules, and masses of the inorganic world. In order to do this, it had called in the aid of the mathematician. By his technical language, alone, can any problem of time and space, matter and motion be exactly and definitely solved. Even in the organic world, he had begun to render the botanist and zoölogist important services. He had shown that, in the plant, there is a law of extreme and mean ratio; that this is the nearest approach to a uniform distribution of the leaves; he had shown that, in the zoölogist's classification of animals, the first great division rests upon laws of mechanical equilibrium in the embryo, as imperious as the law of an arch. The natural expectation was that mathematical bases would in like manner be found for the division of classes, orders, families and genera, if not the species.

But under the theory of continuous gradual development this course of scientific progress seems arrested. From the fact that between any two forms of nature we can always, if we have numerous specimens, find intervening links, it is assumed that there is no real line of demarcation between the forms. There is a double fallacy in this assumption. Between any chestnut and any oak, for example, it may be possible to find an intermediate form. But it does not follow that it would be possible to find a form of which a thoroughly trained botanist would say, "This is either an oak or a chestnut, but I cannot tell which." It may easily be true that no definition, in words, of an oak would exclude every chestnut; nor any definition of a chestnut exclude every oak. It does not follow that the senses would not distinguish; a sharp observer sees many differences not easily to be described in words; as those between an oak and a chestnut, an apple and a pear, a plum and a cherry. The history of science is full of instances in which a more acute observer has learned to see and point out distinctions between things which had been confounded. And, secondly, if it were possible to show that between two groups the transition is so gradual that no eye can detect the dividing line, it would not follow that no dividing line existed, nor that the groups had a common origin. The intellect can sometimes clearly distinguish things, indistinguishable by sense. The elastic curve has, at the extremities of its series of variations, a straight line and a circle; indistinguishable by the eye, or by the imagination, from those produced by varying the eccentricity of the ellipse. Yet reason, transcending imagination, shows them to be entirely different; they cannot be made alike, except by a process which would confound all intellectual distinctions, and destroy the possibility of science.

The ellipse and the elastic curve belong to genera further removed from each other than the oak and the chestnut; yet each may pass by variation into the form of a circle. Similar instances abound in the curves investigated by geometry. An observer unacquainted with mathematics might think it easy to pass the elastic curve into the form of a circle, and then elongate it to an ellipse; or easy to frame a definition of the one curve, which should also include the other; the geometer knows that neither is possible. When the forms of the chestnut and the oak are as thoroughly understood by the botanist, as those of the ellipse and the elastic curve are by the geometer, he will probably wonder that the two genera were ever considered difficult to define and separate. It may even be that the mathematician will demonstrate the difference of the forms. It is the mathematician alone who has introduced precision and certainty into the other physical sciences; and he will probably, at some day, introduce them into biology. The botanist and the zoölogist may rebel, but they will rebel in vain. The numbers of Pythagoras and the axioms of Euclid are inexorable. The fates themselves cannot violate the laws of geometry, arithmetic, and algebra; much less can fluttering theorists break through those adamantine bars.

The erratic genius of De Maillet, and of Erasmus Darwin, has built an ingenious theory, perfected under the magic influence of Charles Darwin, by the aid of a host of writers, which flatters itself that it explains all organic forms in complete independence of geometrical and arithmetical law. It makes the labors of classification as empty of real meaning as though they had been expended upon the forms of clouds or upon the disposition of the settlings of cups of tea. This theory of insensible, accidental variation, modified and arrested by the surroundings, is a virtual assertion that the whole problem of classification is a delusion. Expand the theory to its most complete form, and it becomes self-contradictory and self-destructive. It would declare all personification of objects, and all entification of attributes, mythical and illusory, and make monotheism the last step of the illusion previous to awaking and discovering that it is all a dream. Thus its evolution contains in itself an illogical breach of continuity. When a mathematical series tends through an indefinite series of

THE DEVELOPMENT OF FORMS. 107

terms toward unity, it is illogical to assume either that it ends in zero or that it becomes indefinite. As well might the modern correlation of forces, tending constantly to show that all phenomena are modes of motion, be held to show that there is no motion, or at least that we can not know that there is motion, as the universal tendency of cultivated thought to reduce all supernatural forces to the will of one God be held to lead properly to atheism or to agnosticism. If the reduction of personification in things and ideas to smaller and smaller numbers logically leads to the denial of real being in any entified idea, it should lead also to the denial of personality in any thing; that is, to the denial of personality in our fellow-animals, and even in our fellow-men. In fact, this theory goes further; it ignores the existence of personality in one's self; it makes the theorist himself a non-existent being, knowing only that one thing, that he does not exist.

The students of botany and zoölogy have been laboring for nearly twenty-five centuries in one direction, with steady progress; there has been a substantial agreement among them as to the proper divisions of the animal and vegetable kingdoms; their divisions have been concerning the grounds of the division. For the final settlement of the questions of classification the aid of the mathematics is necessary. This has been the destiny of the other physical sciences; it is that of biology also. There is no breach of continuity in nature; geometric and algebraic law rules absolute over all that can be bounded in space and time. The vagueness of arbitrary variation and survival of the fittest is a poetical dream; it must give way to the intellectual, scientific sternness of invariable law, bounded by invariable limits. As the four primitive forms of the embryo flow from necessary mechanical conditions, inflexible as the law of equilibrium in arches; so the classes, orders, families, genera, will be found to be formed by conditions sharply defined in nature; and hereafter to be, through the aid of mathematics, sharply defined in human thought.

No geometer will willingly relinquish his hope of great triumphs in the future; when the new mathematical methods of the nineteenth century shall have been as faithfully applied to the problems of organic form, as the methods of the seventeenth century have been to those of inorganic matter. If the melodious wail of Darwin's bugle leads the naturalists to retreat from the grand problem of classification; and if the trumpet of the elder Agassiz fails to rally them; the mathematicians will press forward and gain the high honor of victory, on the noblest field of natural science. Mathematical science cannot admit the possibility that the rhythm and symmetry of the organic kingdoms is an accidental result of accidental variations; there must be algebraic and geometric law at the basis, not only of each organic form, but of the series of forms. The series has a unity; capable, when men have attained a fuller comprehension of it, of expression in terms of thought.

The rhythm and harmony of a symphony reveal not only the skill of the orchestra and its conductor; but the great mind and noble heart of the composer. The rhythm and harmony of the organic world reveal the power, the wisdom, and the love of God. So long as man is less than the universe, his wisest and best course is to seek everywhere, not for discords and maladaptations, but for harmonies, correlations, adaptations. The universe is the sum of all symmetries; and contains all geometries, architectures, sculptures, and pictorial arts. It is the sum of all rhythms, melodic or harmonic; and contains all algebra, poetry, music and dance. The Divine Word, which created it, is wisdom and love; and manifests wisdom and love in every syllable and tone in which it utters itself; not least in the wondrous series of the forms of plants and animals; swaying, in the responsive rhythm of growth and decay, sleep and activity, generation and succession, to the periodic march of the earth, the moon, the planets and the sun.

www.ingramcontent.com/pod-product-compliance
Lightning Source LLC
Chambersburg PA
CBHW030905170426
43193CB00009BA/739